Series in Biomedical Engineering

Series Editor: Joachim H. Nagel

Series in Biomedical Engineering

Editor-in-Chief

Prof. Dr. Joachim H. Nagel
Institute of Biomedical Engineering
University of Stuttgart
Seidenstrasse 36
70174 Stuttgart
Germany
E-mail: jn@bmt.uni-stuttgart.de

The International Federation for Medical and Biological Engineering, IFMBE, is a federation of national and transnational organizations representing internationally the interests of medical and biological engineering and sciences. The IFMBE is a non-profit organization fostering the creation, dissemination and application of medical and biological engineering knowledge and the management of technology for improved health and quality of life. Its activities include participation in the formulation of public policy and the dissemination of information through publications and forums. Within the field of medical, clinical, and biological engineering, IFMBE's aims are to encourage research and the application of knowledge, and to disseminate information and promote collaboration. The objectives of the IFMBE are scientific, technological, literary, and educational.

The IFMBE is a WHO accredited NGO covering the full range of biomedical and clinical engineering, healthcare, healthcare technology and management. It is representing through its 58 member societies some 120.000 professionals involved in the various issues of improved health and healthcare delivery.

Previous Editions:

Spaan, J. et al: BIOMED, Biopacemaking, 2007, ISBN 978-3-540-72109-3

Sekihara, K., Nagarajan, S.: BIOMED, Adaptive Spatial Filters for Electromagnetic Brain Imaging, 2008, ISBN 978-3-540-79369-4

Madhavan, G., Oakley, B.; Kun, L.(Eds.): BIOMED, Career Development in Bioengineering and Biotechnology, 2009, ISBN 978-0-387-76494-8

Michel J.A.M. van Putten

Essentials of Neurophysiology

Basic Concepts and Clinical Applications for Scientists and Engineers

Springer

Michel J.A.M. van Putten
Department of Neurology and Clinical Neurophysiology
Medisch Spectrum Twente and
Institute of Technical Medicine
University of Twente
PO Box 217
7500 AE Enschede
Netherlands
E-mail: m.j.a.m.vanputten@tnw.utwente.nl

ISBN 978-3-540-69889-0 e-ISBN 978-3-540-69890-6

DOI 10.1007/978-3-540-69890-6

Series in Biomedical Engineering ISSN 1864-5763

Library of Congress Control Number: 2009921735

Typeset & Cover Design: Scientific Publishing Services Pvt. Ltd., Chennai, India.

Printed in acid-free paper

9 8 7 6 5 4 3 2 1

springer.com

IFMBE

The International Federation for Medical and Biological Engineering (IFMBE) was established in 1959 to provide medical and biological engineering with a vehicle for international collaboration in research and practice of the profession. The Federation has a long history of encouraging and promoting international cooperation and collaboration in the use of science and engineering for improving health and quality of life.

The IFMBE is an organization with membership of national and transnational societies and an International Academy. At present there are 53 national members and 5 transnational members representing a total membership in excess of 120 000 professionals worldwide. An observer category is provided to groups or organizations considering formal affiliation. Personal membership is possible for individuals living in countries without a member society. The IFMBE International Academy includes individuals who have been recognized for their outstanding contributions to biomedical engineering.

Objectives

The objectives of the International Federation for Medical and Biological Engineering are scientific, technological, literary, and educational.Within the field of medical, clinical and biological engineering its aims are to encourage research and the application of knowledge, to disseminate information and promote collaboration.

In pursuit of these aims the Federation engages in the following activities: sponsorship of national and international meetings, publication of official journals, cooperation with other societies and organizations, appointment of commissions on special problems, awarding of prizes and distinctions, establishment of professional standards and ethics within the field as well as other activities which in the opinion of the General Assembly or the Administrative Council would further the cause of medical, clinical or biological engineering. It promotes the formation of regional, national, international or specialized societies, groups or boards, the coordination of bibliographic or informational services and the improvement of standards in terminology, equipment, methods and safety practices, and the delivery of health care.

The Federation works to promote improved communication and understanding in the world community of engineering, medicine and biology.

Activities

Publications of the IFMBE include: the journal Medical and Biological Engineering and Computing, the electronic magazine IFMBE News, and the Book Series on Biomedical Engineering. In cooperation with its international and regional conferences, IFMBE also publishes the IFMBE Proceedings Series. All publications of the IFMBE are published by Springer Verlag.

Every three years the IFMBE hosts a World Congress on Medical Physics and Biomedical Engineering in cooperation with the IOMP and the IUPESM. In addition, annual, milestone and regional conferences are organized in different regions of the world, such as Asia Pacific, Europe, the Nordic-Baltic and Mediterranean regions, Africa and Latin America.

The administrative council of the IFMBE meets once a year and is the steering body for the IFMBE. The council is subject to the rulings of the General Assembly, which meets every three years.

Information on the activities of the IFMBE are found on its web site at: http://www.ifmbe.org.

To my parents who learned me
to see, enjoy, explore,
and share my fascination
with Science & Technology

Preface

In this book, we approach neurophysiology at the interface of neurology and clinical neurophysiology. The medical disciplines of the nervous system, neurology and clinical neurophysiology, rest heavily on other sciences, notably cellular biology, neuro-anatomy, neuro-physiology, applied physics and mathematical biology.

Existing medical textbooks on neurophysiology, neurology and clinical neurophysiology are an excellent source of the phenomenology of various principles and diseases. Here, we choose to elucidate some of the underlying physiological, physical processes and experimental methods, intended for a broad audience – medical residents and students, as well as students in the emerging area of medical technical sciences.

We feel that a good understanding of fundamentals may significantly enhance insight into various aspects of clinical neurology and clinical neurophysiology. This book, therefore, is focused on a selection of clinical signs and symptoms to highlight *basic principles of neurology, (neuro-)physiology and neuroanatomy.*

While we believe this text to be of interest to medical students or residents in neurology or clinical neurophysiology, we specifically aim at students interested in contributing to new developments and innovations in neurology and clinical neurophysiology. These students are involved with patients, even though they are not trained for routine patient care.

Exciting new developments in neurophysiology are brain computer interfaces, artificial neurons, artificial nerves, deep brain stimulation, advanced signal processing for brain monitoring, development of new (neuronal) tissues, medical decision making, and the (neurocomputational) modeling of diseases like Parkinson or epilepsy. With our selection of topics, we hope to bring the reader up-to-date to some of these promising frontier topics.

This book is based on lectures given at the University of Twente for advanced undergraduate students from the Institute of Technical Medicine. Throughout the text, we try to provide a balanced combination of diverse, but related, topics. We will deal with both anatomy and physiology, but, when it is helpful, concepts from mathematics or physics are included in our discussion along the way.

This book, therefore, presents a hybrid of information, both from the clinical field and from basic physiology, with an emphasis on concepts. Where needed, mathematical descriptions are included for a quantitative understanding. We include prototype examples of neurological diseases. In this approach, we address the more technically or research minded students or residents, as well.

The topics covered range from basic neuroanatomy to physiology, from membrane potentials to interneuronal communication. In addition, we discuss how various measurements on the human nervous system can be performed, including a treatise on medical measurements in general. A short chapter on medical decision making is included, since this is a potential field for innovation, as well (computer supported decision making). Throughout the text, we will present various examples of neurological diseases.

After the introductory Chapter, Chapters 2-4 discuss the concepts of excitable cells and neural circuits. Chapter 5 is devoted to basic measurement techniques, with a strong emphasis on engineering and physics principles. In Chapter 6, we discuss how we can electromagnetically excite neuronal cells. This is relevant for both diagnostic as therapeutic purposes. Chapters 7 & 8 present essentials of neurophysiological measurements (EEG and EMG), the field of clinical neurophysiology, and its role in the diagnostic process. Chapter 9 is a short introduction to digital signal analysis, with an emphasis on extracting some quantitative features from the human EEG. In Chapter 10, we present a few new and exciting therapies, such as deep brain stimulation, vagus nerve stimulation for the treatment of some forms of epilepsy and brain computer interfaces. Finally, Chapter 11 discusses some concepts of information as relevant for diagnostic strategies, including various elements of "rational decision making".

I am very happy I met Christoph Baumann from Springer Verlag at the Medicon 2007: XI Mediterranean Conference on Medical and Biological Engineering and Computing in Ljubljana, Slovenia. Here, we discussed the possibility to publish the Lecture Notes as a textbook, and I was pleased that Springer agreed to do so. Soon after, I obtained very useful suggestions and comments from one of Springer's reviewers, that has contributed to various improvements in the text. I further gratefully acknowledge Susan van Maarseveen, Bennie ten Haken, Meyke Roosink, Hanneke van Genderen and Daphne de Klerk for critically reading the manuscript. In addition, several students have made suggestions and pointed out previous errors and mistakes in the Lecture Notes, upon which this book is based. In particular, I would like to thank Niels Langhout, Willemijn van der Kooi, Marjanne Ymker, Erik van Bussel, Jasper Jochem, Daan de Jong and Tianne Numan. My colleague Jeroen van Vugt pointed out various inconsistencies regarding my treatise on the basal ganglia and Parkinson's disease. My brother, Maurice, is acknowledged for his most valuable feedback on the Chapters 1-3. Many thanks also to Sid Visser, from the department of applied mathematics, who not only made many of the illustrations, but also assisted in the critical reading of the manuscript. I thank Stephan van Gils for his very careful reading of the Chapters 2 and 3, that led to various improvements. Last but not least: Regina, thank you for your understanding, critical comments, and support.

Enschede
January 2009

Michel J.A.M. van Putten

Contents

Part I
Basic Neurophysiology

Chapter 1
Introduction

Many of our unique human characteristics arise from our central nervous system. The brain allows us to experience happiness, sadness or pleasure. Thanks to this wonderful organ we can laugh, think, make plans, and move around in the world. Almost all interactions with our environment further involve our motor system, that is controlled by our central nervous system, as well, through the peripheral nerves. Due to this intricate relationship between the central and peripheral nervous system and the motor system, both belong to the domain of neurology and (clinical) neurophysiology.

The central nervous system consists of a huge amount of very specialized cells: neurons and glia cells. The neurons are the basic structural components of the brain, that are specialized to process information. Individual neurons can generate fast changes of the membrane voltage, by controlling selective movements of ions across their semipermeable membranes. In this way, information that is processed by a neuron is communicated with a particular electro-chemical signature[1].

Neurons communicate continuously with other neurons, both sending and receiving signals, that are typically conducted between the various neurons by chemical synapses or gap-junctions. Rodolfo Llinás, a noted neuroscientist from the New York University School of Medicine, phrased this process once as follows: "Neurons like one other very much. They respond to one another's messages, so they basically chat all day, like people do in society". In this way, functional neural networks are created, often dedicated to a particular process, for instance language, perception or memory. Our brain is a large assembly of these networks.

In addition to neurons, the other highly relevant cell types in the brain are the glia cells. These cells partially have a supportive function, but are relevant in information processing, too. This includes e.g. the buffering of various ions that are essential for proper functioning of the neurons (as we will discuss later), e.g. the extracellular potassium.

Some of the frontier questions in neurophysiology are: What is the relation from ion movements across cell membranes to memory, perception, movement

[1] At present, it is not clear (or proven) that all information states and processes associated with neuronal activity is encoded exclusively in the electro-chemical domain.

M.J.A.M. van Putten: Essentials of Neurophysiology, BIOMED, pp. 3–5.
springerlink.com

or creativity? What are the basic properties of neuronal communication? How is information stored in the nervous system? How is the desire to move put into action, and transmitted to a particular muscle? What is the role of brain rhythms in relation to the functioning of neurons, neural networks and their interactions?

Although the nervous system typically works extremely well to control and process information, there are many disturbances, collectively known as neurological or psychiatric diseases. Examples include epilepsy, stroke, Alzheimer's disease, Parkinson's disease, migraine, and various neuromuscular disorders, such as neuropathies or amyotrophic lateral sclerosis (ALS)[2].

Neurology and clinical neurophysiology study and treat these derangements or diseases of the nervous system. In addition to the central nervous system disorders, neurology and clinical neurophysiology are concerned with abnormal functioning of the peripheral nervous system and the muscles.

Neurology, neuroscience, and clinical neurophysiology are relatively new medical fields. Neurology as an independent medical specialty started in 1882, when Jean-Martin Charcot (1825-1893) was appointed a chair in Paris. He was the first "Professor in Clinical Neurology" to investigate "diseases of the nervous system". The Paris school of Charcot strongly influenced neurology in the 19-th century. In the Netherlands, it was only about 25 years ago that Neurology became established as an independent medical specialty. Up to that time, neurological patients were typically treated by (neuro-) psychiatrists.

Various methods to categorize neurological diseases are possible, including categorizations based on the cause (e.g. infection, degenerative, trauma, cancer, etc.), the age of occurrence or the particular cells involved. An alternative division of various diseases is a categorization into diseases of the sensors, diseases of the actuators and diseases of the Central Nervous System (CNS). This is illustrated in Figure 1.1.

Tremendous progress has been made in the last few decades. A few decades ago, the neurologist had only very limited therapeutic options; this has changed dramatically. For instance, besides more powerful medication for the treatment of seizures in patients suffering from epilepsy, epilepsy surgery has proven to be a very useful alternative in some patients. Also, electrical stimulation of the central or peripheral nervous system is increasingly used for the treatment of neurological disorders. Placing electrodes in the brain to electrically stimulate specific structures may alleviate various symptoms, e.g. in patients suffering from Parkinson's disease. Electrical stimulation of one of the cranial nerves, specifically the vagus nerve, is increasingly used to reduce the seizure frequency in patients with epilepsy. Interestingly, the detailed working mechanism of these therapies is still not known. This is a challenging

[2] In the United States, ALS is often called Lou Gehrig's disease because of New York Yankees' star Lou Gehrig, who was diagnosed with ALS in the 1930s. People in England and Australia call ALS "Motor Neurone Disease" (MND). The French refer to it as "Maladie de Charcot", after the French doctor Jean-Martin Charcot, who first wrote about ALS in 1869.

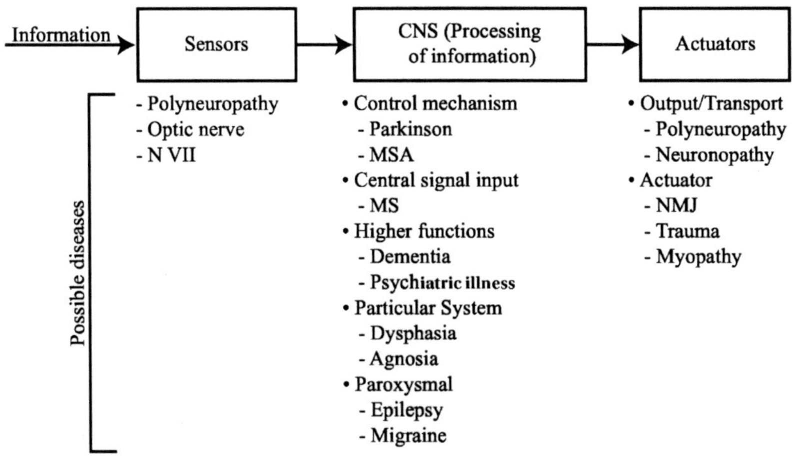

Fig. 1.1 Overview of various neurological diseases. Input diseases, output diseases and disorders of the central nervous system (CNS). Disorders of the output comprise transport disorders (peripheral neuropathies), diseases of the actual actuators, e.g. myopathies, or the neuromuscular transmission. Although processing of information is not limited to the central nervous system (one can argue that processing already takes place at the various sensors, e.g. by coding the strength of a stimulus to a particular frequency of action potentials), the complexity of the processing that is present in the central nervous system is orders of magnitude larger than in any other part of the human body. MS=multiple sclerosis; NMJ=neuromuscular junction and MSA=multiple system atrophy.

research field, where many new applications are foreseen, as well[3]. Disease modifying drugs (e.g. interferon in multiple sclerosis) are available for various neurological diseases. Perhaps, in the future, artificial neurons and artificial "brain modules" can improve the life of patients.

Besides therapeutic improvements, a tremendous improvement in diagnostics has been realized, as well, in particular due to advanced imaging with Magnetic Resonance Imaging (MRI), functional MRI, Positron Emission Tomography (PET), single photon emission computed tomography (SPECT) and magnetoencephalography (MEG). While not so long ago, patients underwent a rather painful pneumoencephalography, where air was injected into the ventricles of the brain to try to diagnose e.g. a brain tumor, we now have scanners available, that allow us to make detailed images of the brain without (almost) any discomfort to the patient.

This short sketch of improvements and innovations shows that the medical sciences of the nervous system are very dynamic fields, and many other improvements are expected in the near future. Students with a keen interest in this field who wish to further these innovations will have a most promising future ahead.

[3] Examples include the deep brain stimulation for the treatment of mood disorders (depression), adipositas and various pain syndromes.

Chapter 2
Excitable Cells and Action Potentials

2.1 Introduction

The function of the central and peripheral nervous system is *to process information.* This information is abundant in the outside world, and includes written information (newspapers, textbooks), sounds & vision (MTV), smell, etc. Not only information from the outside world is processed by the nervous system, but also information from within the organism, e.g. for controlling the heart rate or respiration, using the signals from various sensors inside the body[1]. Additionally, we are also able to somehow *create* new information by combining stored knowledge (sometimes with actual real-time new input): this process is known as "thinking" or "creativity". For this introduction, we will not go into further detail, but it is clear that the nervous system is able to process information from various sources and for various purposes, where these processes may occur both consciously and unconsciously, and both simultaneously and sequentially. This is schematically shown in Figure 2.1.

In the current context, we could define "information" as all those signals that have a particular meaning for the system, and subsequently can potentially influence the behavior of the system at hand. It is clear that the information is carried by a particular form of energy, e.g. electromagnetic radiation or mechanical energy. An example of the latter category is sound, that may contain information "coded as" spoken language. In order for the organism to process these signals, it is typically necessary to *transform* these signals to a different energy domain. In your computer, the energy that transports the actual information is electrical energy (electrons). In the human body,

[1] An example that can easily be tested is the awareness of position using our position sensors. If you close your eyes, and have someone else put your arm in a certain position, you know this position without having to open your eyes. This information is obtained from the position sensors, and you are perfectly able to put your other arm in the same (mirror) position, still with your eyes closed.

M.J.A.M. van Putten: Essentials of Neurophysiology, BIOMED, pp. 7–32.
springerlink.com

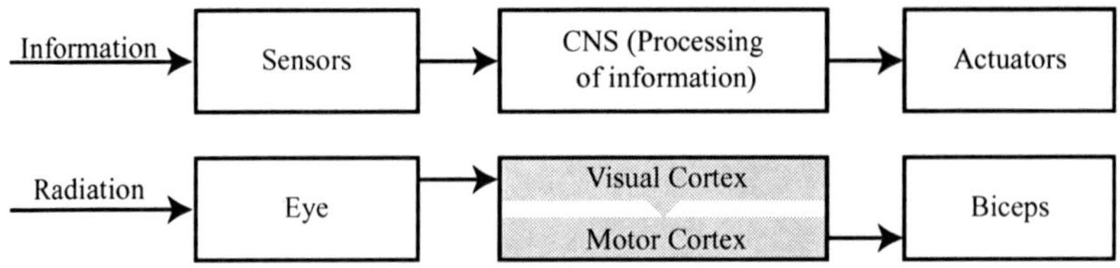

Fig. 2.1 Schematic of the central and peripheral nervous system with the input and output relations as an information processing system with input and output relations. The sensor is not only involved in the recording of incoming information, but also in the conversion to a particular energy domain. Radiation is converted by the rods and cones in the retina to electrochemical energy, that is subsequently send to the visual cortex and other brain areas involved in visual perception. A similar process occurs at the actuator side. In this example, the actuator is a mechanical device: the biceps muscle.

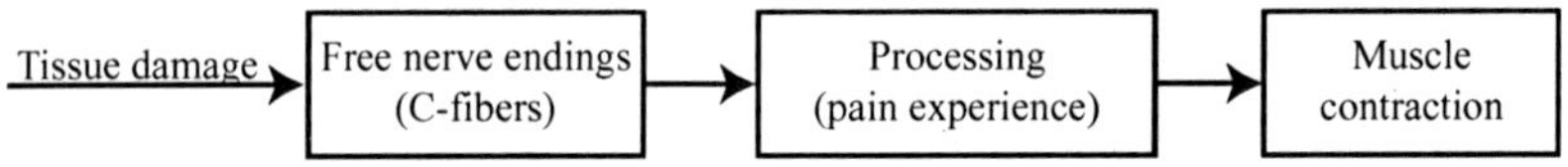

Fig. 2.2 Schematic of processing of tissue damage, conveyed by sensory fibers.

the energy for the actual processing (as we know it) is of an electrochemical form, carried by action potentials and neurotransmitters[2]. However, the energy from the outside world is often from a different domain: mechanical (sound, touch), electromagnetic (vision), etc. Fortunately, we have several transducers to transform these different forms of energy to the electrochemical domain[3]. As an example, visual input carried by electromagnetic radiation is converted by nerve cells in the eye to action potentials, i.e. electrochemical energy. For this conversion, a very special transducer or *sensor* has been "developed". Once this conversion process has taken place, the information is transported in the organism to our central processing unit: the brain. From there, usually a particular action is taken: the brain on its turn may send output signals to an output transducer, which is known as an *actuator*. For animals and man, the most relevant actuators are the muscles[4].

As an additional example of these concepts, Figure 2.2 shows a schematic of the flow of information that occurs in the case of tissue damage. If this happens, for instance due to an inflammatory process in your skin, particular

[2] At present, we are ignorant of more exotic processes, such as quantum mechanical neurological activities, if any. While of potential interest, perhaps to consciousness, these possibilities are speculative at present as we have no experimental methods at hand to probe this arena.

[3] Do you know some examples of organisms that have sensors for a very limited number of input energies, only?

[4] This holds for actuators that relate to the outside world. For homeostatic control, a variety of alternative actuators are present (internal organs, like the kidneys, for example).

nerve fibers, that are most sensitive for e.g. the chemicals that are generated during an inflammation start to send action potentials to the brain. In this sense, these nerve fibers with their specialized endings serve as the input transducer, i.e. the sensor. Note, that in this case, the actual input information is the concentration of particular substances in the skin. The action potentials finally arrive in particular brain structures, and the central processing unit may decide for some action, e.g. to contract your leg muscles to go to your bathroom to look for an aspirin, or perhaps to call your physician.

Another example illustrates the flow of information for a visual input, for instance you are walking to the super market, and your friend John appears to be there, as well. This is noted because photons from his face, arranged in particular order, are captured by your eye, and projected upside down on your retina, where numerous rods and cones start generating action potentials. At this point, a conversion from electromagnetic energy to, again, electrochemical energy, has occurred. The action potentials will travel through the numerous axons of the optic nerve to both hemispheres, and finally arrive in the occipital lobes. Here, other neurons process this information, and send out action potentials to various brain areas. If, somehow, this particular pattern matches a previous stored image, you may be able to be aware that these photons were actually coming from the face of John[5]. Then, you may decide to send out signals to your muscles of your tongue and mouth, but also to your language and motor area in your brain, to be able to make a sound, to say: "Hi, John". This latter process is a realization of controlling one of your actuators, in this case the muscles of your mouth (and lungs and vocal cords) to make sound: a form of mechanical energy.

The brain has to perform a wealth of different functions, for which different neural circuits are present. This not only includes neural assemblies for performing mathematics or language, but also circuitry to control autonomous

[5] This is an extremely simplified picture of reality. Processing of visual information is a very complicated procedure. Recognition of faces is even more special, and the brain has dedicated modules for face recognition. One of these modules is located at the planum temporale of the dominant temporal lobe (in most case, at the left side). Damage to this area may cause *prosopagnosia*: a selected brain disorder, where patients cannot recognize faces (from visual information) but are otherwise functioning completely normal. An interesting variant to this condition, is known as Capgras' syndrome, which is an example of a disconnection syndrome [1]. This is brilliantly described by Vilayanur Ramachandran, an Indian-American neurologist, in Phantoms of the Brain [2]. In this book, he writes that "the Capgras' patient will typically identify people close to them as being imposters - identical in every possible way, but identical replicas. Classically, the patient will accept living with these imposters but will secretly "know" that they are not the people they claim to be. When Capgras' occurs secondary to other disorders such as the schizophrenias, the experience can take on a frightening dimension along the lines of "Invasion of The Body snatchers". Neurologically speaking, Capgras' Syndrome occurs when the kinesthetic component of facial recognition is damaged. The result being that the strategy of facial recognition no longer has the internal kinesthetic on which to exit. Thus the patient gets caught up in an eternal loop in his strategy and needs to form an explanation (the delusion) in order to understand why this experience should occur."

functions, such as blood pressure, temperature, respiration and growth. In addition, there are many circuits devoted to the control of muscle activity, necessary for most of our communication with the outside world, including locomotion.

Although these various systems in our brain are very efficient at performing their individual tasks, they are limited in flexibility. For instance, the motor cortex does not appear able to perform visual functions[6]. This is presumably not because the motor cortex contains cells that are essentially "motor" in what they do but rather it is because of their morphological specialization and the particular network they are making part of. In fact, the basic architecture of the cortex is very similar. There are six main layers, comprised of pyramidal cells and various interneurons[7]. These cortical micro-circuits are functionally regrouped into cortical columns. These columns can be regarded as basic functional units. The particular function of neurons, therefore, is determined by the neuron's computational role in a *function-specific neural network* [4].

How exactly the information is coded in the nervous system, however, is still far from understood. We do know, that the information at the neuron level is contained in the particular train of action potentials. How this information is integrated at the network level with all the input stimuli from other neurons, is only partially clarified, and a topic of intensive research. The brain ultimately succeeds in making a coherent and conscious percept of all the information that is perceived by the various senses. It is an intriguing scientific question how this is realized by the brain. This is known as the "binding problem".

We will now proceed with a more detailed analysis of the most important basic element of the nervous system, the neuron itself. As we will find out, this is indeed a truly remarkable entity, that, by cooperation with many other neurons, allows us to think, play, enjoy and be aware.

2.2 The Neuron

In the human body, information transfer is by electrochemical signals, mediated by specialized cells. These cells are dedicated to the generation and transport of information, and are present in the central and peripheral nervous system, the muscular system and some cells of the endocrine system.

[6] Although it is involved in processing of visual information. Recently, observations in monkeys (Macaque) have shown that seeing a particular motor action can result in similar firing behavior of a single neuron as performing this motor action. Cells with this property are called "mirror neurons", and most likely exist in man, as well [3].

[7] We will discuss this in some more detail in a later stage. Here it suffices to mention that if we examine microscopically a part of the cortex of the brain, different layers are observed, that run in parallel to the cortical surface. These layers have distinct morphological properties, that are caused by both differences in the density of neurons and the particular neurons present. As you may know from your neuroanatomy course, the nervous system is comprised of various neuronal types, for instance pyramidal cells, stellate cells and chandelier cells.

They are collectively known as *excitable cells*. By definition, these cells show a *functional activity* after a particular stimulation. In rest, these cells are characterized by a *rest membrane potential* of about -90 to -60 mV across their cell membrane, which changes to about +50 to +60 mV in the generation of an *action potential*. In the central nervous system itself, the neuron has been taken as an elementary unit of processing. Cajal's work in the late nineteenth century provided a strong basis for this insight. He argued that neurons are distinct, individual cells, separated by each other by a spatial gap[8], and predicted the existence of an additional intracellular mechanism for communication from one neuron to another. It is known now that neurons receive input from other neurons via their dendrites, and pass information via their long output channel, the axon, that may branch several times, connecting to other neurons, glands or muscles.

Various different neurons exist. Already in the retina, which is actually a part of the central nervous system, we find cell types with different morphologies, sizes and connectivity patterns. In addition, there exist many functional and biochemical differences between neurons, including their energy requirements. Cortical pyramidal cells, for instance, have quite distinct computational properties compared with the various inhibitory cells in the neocortex. Notwithstanding this diversity in neurons, there are common features. Firstly, like any other cell in the human body, nerve cells have a membrane, that surrounds the cytoplasm that contains the nucleus, mitochondria, nucleoli. The basic anatomy is shown in Figure 2.3.

Secondly, all neurons exhibit a resting membrane potential and are capable of generating action potentials. These action potentials - "self-generated" temporary changes in the cell membrane potential- can propagate along the axons. It is considered to be the most basic level of information processing. The interface between a terminal axon branch and another neuron is known as the synapse, illustrated in Figure 2.4.

At the synapse, an axon releases a *neurotransmitter* in response to an action potential, that diffuses through the spatial gap, the synaptic cleft, to activate a particular receptor, located at the receiving dendrite. Most synapses in the central nervous system are *chemical synapses*. These can be viewed as rectifiers: a signal can be transported from the terminal axon to the dendrite, but not vice versa. In addition, there is a time lag between arrival of the signal at the terminal axon and the response of the dendrite of the order of 1 ms. Besides chemical synapses, the CNS contains electrical synapses, so called gap junctions. Here, cell membranes of different neurons are intimately

[8] This spatial gap is now known as the synaptic cleft. Besides the separation by gaps (where neurotransmitters play an important role, as discovered later), there are structures where cells are physically joined by conduction "gap junctions". Here, cells approach within about 3.5 nm of each other, a much shorter distance than the 20 to 40 nm distance that separates cells at chemical synapses, and communication is bi-directional. These electrical synapses are faster and more reliable than chemical transmission, but are also much more limited in flexibility.

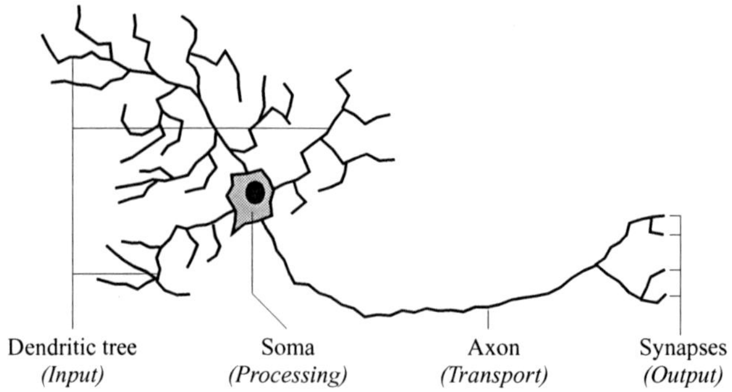

Fig. 2.3 Cartoon of a neuron. There are various dendrites, transporting information towards the cell, and a single axon, that transports information from the cell. Typically, an axon branches at its end, where it may end on different neurons. This interface is known as a synapse.

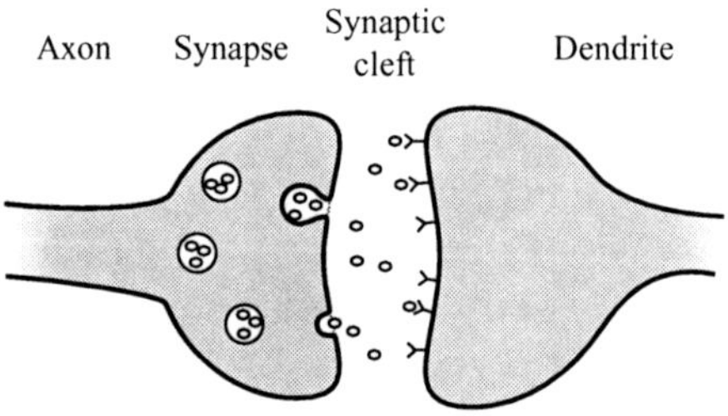

Fig. 2.4 Shown is a chemical synapse. Information flow is from left to right.

connected, and signal transmission is by direct electrical communication. At the gap junction, there is virtually no time delay present. More details about the processes that occur at synapses will be treated in a chapter 4. In the remainder of this chapter, we will focus on the generation of the cell-membrane potential and the generation of action potentials. Conduction of the action potentials is discussed in chapter 3.

2.3 Generation of the Cell-Membrane Potential

All neurons are surrounded by a semipermeable or selective membrane, which consists of a vast amount of fats and proteins. This is schematically illustrated in Figure 2.5. This cell-membrane separates the fluid inside the cell, the *intracellular fluid* from the fluid outside the cell, the *extracellular fluid.* The extracellular fluid provides the cell with energy and creates the right environment (milieu exterieur) for the cell.

The constitution of the intra- and extracellular fluids is quite distinct, as is shown in Table 2.1. There are different mechanisms that play a role in the

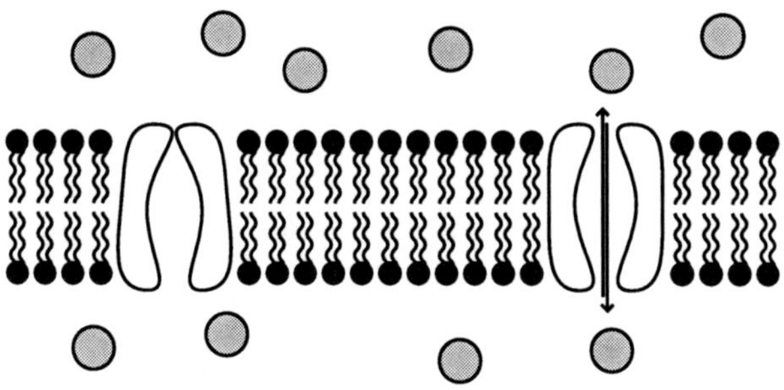

Fig. 2.5 Cartoon of the cell membrane. The lipid bilayer with the ion channels functions as a semi-permeable membrane, that separates the interior from the exterior environment.

Table 2.1 Constitution of the intra- and extracellular fluid.

ion or molecule	intracellular mmol/l	extracellular mmol/l
Na^+	10-15	135-145
K^+	135-145	3-5
Ca^{2+}	<1	4-5
Mg^{2+}	50-60	1-3
Cl^-	4	120
HCO_3^-	10	24
phosphates	75	4
SO_4^-	2	1
proteins, amino acids, etc.	99	1

generation of the difference in the constituents of the intra- and extracellular fluid. This will not be discussed in detail; however, the main "generator" of this difference is the sodium-potassium-pump. This is a special pump, located in the cell membrane, that pumps sodium outside the cell, and pumps potassium into the cell[9]. The energy that is required for this pump is provided by ATP, an energy rich molecule, that can be converted to ADP plus energy[10].

Equilibrium between electrostatic forces (forces on charges due to electrical fields) and diffusion forces (fictitious forces on charges due to a concentration

[9] It pumps three sodium ions out of the cell for every two potassium ions pumped in. The pump was discovered in the 1950's by a Danish scientist Jens Christian Skou; he was awarded the Nobel prize in Chemistry, (together with Paul D. Boyer and John E. Walker) in 1997. The Nobel Lecture (The identification of the sodium-potassium pump) can be found at `nobelprize.org/nobel_prizes/chemistry/laureates/1997/skou-lecture.pdf`, and is a fascinating account about the search for biological mechanisms.

[10] Some toxic substances specifically destroy this pump, thereby interfering with the primary mechanism that establishes the cell membrane potential.

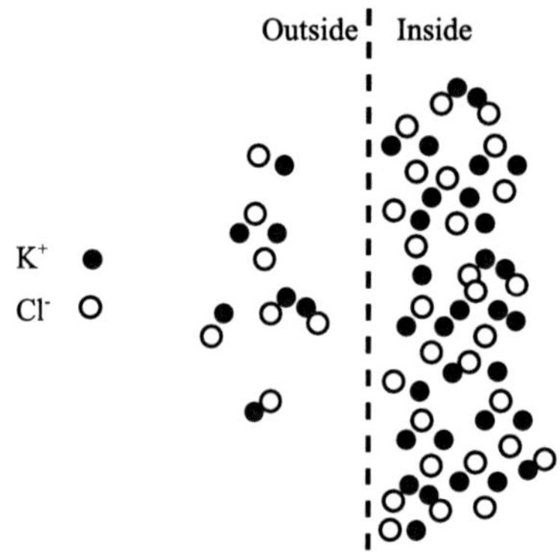

Fig. 2.6 Configuration of a semi-permeable membrane separating two ionic solutions (KCl) with different concentrations on each side of the membrane. Drawing not to scale.

gradient) plays a major role in the generation of the membrane potential and action potentials. It holds in general that, if two compartments with a different chemical constitution are separated by a semi-permeable membrane, such that selective exchange of charged particles is possible, a voltage difference will arise across the boundary of these two compartments. The magnitude of this voltage difference is set by a balance between thermal diffusion and electrostatic drift, by which it will be temperature dependent. The predominant species in the generation of the membrane potential in nerve cells are the sodium, potassium and chloride ions. Before discussing these various ions and their role in the generation of the membrane potential, we will first describe in general how a potential difference can arise across a semi-permeable membrane.

Let us turn our attention to the situation sketched in Figure 2.6. This shows two ionic solutions, containing different concentrations, for instance two solutions of KCl, separated by a semi-permeable membrane.

Let us further assume that the membrane is permeable to potassium (K^+) only, and that the concentration gradient of ions across the cell-membrane is constant. Two forces generally act simultaneously on the charged particles: diffusion forces and electrical forces. Both forces induce movements of the ions involved, thus creating a diffuse flux, J_{diff} and a drift current, J_{drift}. These two currents tend to develop towards an equilibrium. We find for the total flux

$$J_{\text{total}} = J_{\text{diff}} + J_{\text{drift}}. \tag{2.1}$$

Let us now first quantify the diffusion flux J_{diff}, induced by the concentration gradient of the ion species across the semipermeable membrane. Since in this example we are dealing with potassium ions, we write for the diffusion flux of the potassium ion, J_k, according to Ficks first law, that

$$J_k(\text{diffusion}) = -D\frac{d[K^+]}{dx} \tag{2.2}$$

where D is the diffusion constant[11] with unit m^2s^{-1}. The one dimensional flux is of dimension mol $m^{-2}s^{-1}$.

The second force that acts on the ions, is an electrical force. In the presence of an electrostatic field, $E = -dV/dx$, the induced ionic or *drift* current satisfies

$$J_k(\text{drift}) = -\mu Z[K^+]\frac{dV}{dx} \tag{2.3}$$

with μ the electrical mobility ($m^2s^{-1}V^{-1}$), defined as the drift velocity per unit field strength, where Z denotes the valence of the ions. Note that Z is positive for positively charged ions (e.g. $Z = 1$ for Na^+ and $Z = 2$ for Ca^{2+}) and negative for negatively charged ions (e.g. $Z = -1$ for Cl^-).

For the total ionic current it now holds, by substitution in Eq. 2.1, that

$$J_k(\text{total}) = -D\frac{d[K^+]}{dx} - \mu Z[K^+]\frac{dV}{dx}. \tag{2.4}$$

During this diffusive process, the positively charged potassium ions tend to drag along with them negatively charged ions (anions, Cl^-) to assure electrical neutrality in the solutions on each side of the membrane. However, as the membrane is semipermeable, the anions are blocked from accompanying the potassium ions as they move across. This leads to a situation where there are negatively charged ions on one side of the cell membrane and positively charged ions on the opposite side (cf. Figure 2.7).

This actually realizes a third compartment (which as a whole is electrically neutral), with boundaries in the order of the width of several ion radii only. This third compartment consists of a fluid layer on one side of the cell membrane, with anions intracellularly, the cell membrane, and an extracellular fluid layer with the potassium ions. In reality, the thickness of these layers is about 0.8 nm. Note, that electroneutrality[12] is thus preserved, because the charges that are present in this thin layer reside on the membrane, and are not within the solution.

[11] The diffusion constant for a gas is given by $D = lv/3$, with l the mean free path length and v the average molecular velocity. For ions in a fluid, the situation is more complicated, due to the interactions between the ions.

[12] As you remember from your physics course: in any region within an electrolyte, it is expected that the concentration of anions equals that of cations. This condition is known as electroneutrality. Electroneutrality is expected since any net charge in a solution generates very strong electrostatic forces that tend to restore the zero net charge condition.

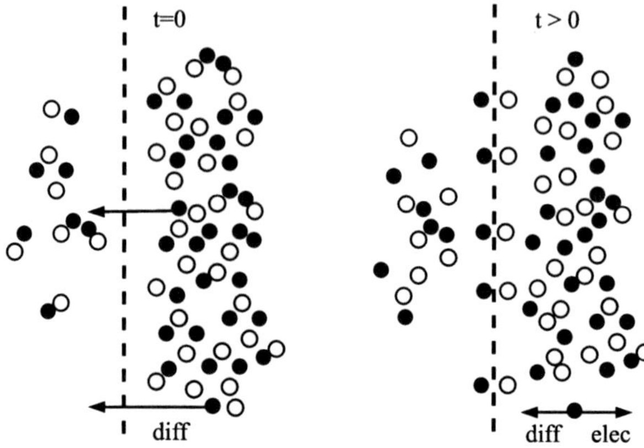

Fig. 2.7 Illustration of the electrostatic and the diffusion forces that act across a semi-permeable membrane. Initially, at $t = 0$, there is a concentration gradient, only. When diffusion starts, charge separation will occur, resulting in a nett electrical force, as well.

Under stationary conditions, where there exists equilibrium between the electrostatic and diffusive forces, we have $J_{\mathrm{k}}(\text{total}) = 0$, and therefore

$$-\frac{kT}{q}\mu\frac{d[K^+]}{dx} = \mu Z[K^+]\frac{dV}{dx} \tag{2.5}$$

where we used the Einstein relationship, that the diffusion constant in the presence of an electrical field is given by $D = \mu\frac{kT}{q}$ where $k = 1.38 \cdot 10^{-23}$ J K^{-1} denotes the Boltzmann constant and $q = 1.6 \cdot 10^{-19}$ C denotes the elementary charge. In some other texts, you may find that $D = \frac{RT}{F}$, where $R = 8.314$ J K^{-1} mol^{-1} denotes the universal gas constant and $F = 96485$ C mol^{-1} denotes Faraday's constant, and T the absolute temperature, which is identical. Prove this yourself. Physical constants can be found in the appendix.

Putting the valence $Z = 1$, we obtain

$$\frac{dV}{dx} = -\frac{kT}{q[K^+]}\frac{d[K^+]}{dx}. \tag{2.6}$$

We can now integrate across the membrane from a place x_0 in the outside compartment, to a place x_i in the inside compartment, where V_o and V_i are the voltages outside and inside the membrane. The potassium concentrations outside and inside the membrane are $[K^+]_o$ and $[K^+]_i$, respectively. Thus we have

$$\int_{V_o}^{V_i} dV = -\left(\frac{kT}{q}\right)\int_{[K^+]_o}^{[K^+]_i}\frac{d[K^+]}{[K^+]}. \tag{2.7}$$

The result is

$$\Delta V = V_i - V_o = -\frac{kT}{q}\ln\frac{[K^+]_i}{[K^+]_o} \tag{2.8}$$

where the transmembrane potential has been defined as the intracellular voltage V_i minus the extracellular voltage V_o.

This equation is known as the Nernst equation. At 37 °C, the term KT/q has a value of approximately 26 mV. Often, however, people use the logarithm base 10, in which case the Nernst equation is multiplied by the factor 2.303. Thus the Nernst equation may be stated at 37°C to approximate to

$$\Delta V = V_i - V_o = -61\log_{10}\frac{[K^+]_i}{[K^+]_o} \tag{2.9}$$

with ΔV in mV. This potential difference is known as the Nernst potential. It quantifies the voltage difference during equilibrium across a semipermeable membrane, where we deal with a single ion species, only. The Nernst equation, therefore, defines the Nernst potential for a given single ion as a function of the concentration gradient, temperature and valence, according to Eq. 2.8.

The voltage difference that we assumed to be applied is realized by the movement of the ions, that occurs by the diffuse forces, set up by the concentration gradient. So, the diffusion process itself creates the voltage difference. If an equilibrium exists between the electrostatic and the diffusive forces, then the diffusion will continue until Nernst equation, Eq. 2.9, is satisfied. In the equilibrium situation it holds that $J_{\mathrm{k,drift}} + J_{\mathrm{k,diff}} = J_{\mathrm{k}}(\mathrm{total}) = 0$. This equilibrium is known as the Donnan equilibrium. This description of the cell membrane potential assumes a static equilibrium, since in the derivation of Nernst equation we put $J_{\mathrm{k}}(\mathrm{total}) = 0$. For completeness, we mention that the Nernst equation, or variants of this equation, always applies if two conducting media are brought into contact. The name of the voltage difference that will arise depends on the kind of boundary between these two conducting media and/or the physical properties of the conducting media. In Table 2.2 some of these potential differences are presented.

If more ion species are present in both compartments, separated by a semipermeable membrane that allows the passage of more than one ion species, we need another description of the membrane potential. In these circumstances,

Table 2.2 The different names of the potential differences that may arise if two different conducting media are brought into contact.

Media/Boundary	Potential Difference
semipermeable membrane	membrane potential
electrolyte-electrolyte	liquid-junction potential
metal-metal	contact potential
semiconductor-semiconductor (pn-junction)	contact potential
metal-electrolyte	half-cell potential

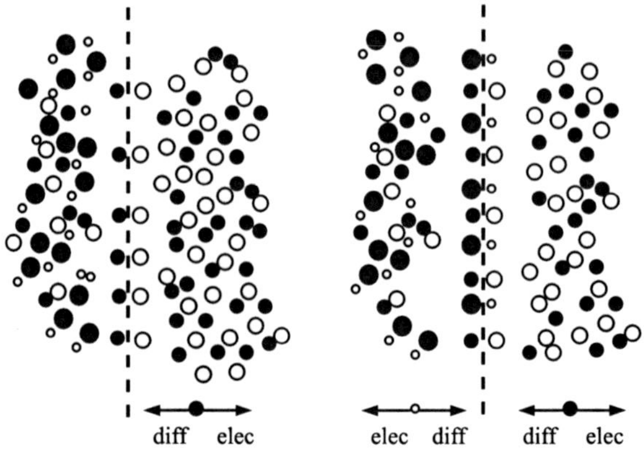

Fig. 2.8 Illustration of a semipermeable membrane separation different ionic solutions. At the left, the membrane is semi-permeable for one ion species, only (illustrated by the small, black dots). At the right, the membrane is semi-permeable for two ions (small black dots and the small open circles), as indicated. The equilibrium potential that is present in this latter situation needs to be described by the GHK-equation. This is a quasi-static equilibrium, since the *sum* of the ionic currents is zero, but the individual ionic currents do not need to be (and are typically not) equal to zero.

it holds during equilibrium that the sum of all currents due to an ion species i (either diffusion or drift) is zero, which we can write as $\sum J_i = 0$. This does not imply, however, that for each current it holds that $J_i = 0$. This situation is called a *quasi-static equilibrium* (cf Figure 2.8). This is a more realistic situation, since in real-life, the membrane is not only semipermeable for potassium ions, but also for chloride and sodium ions. A typical rest membrane potential is about -70 mV, which is close to the Nernst potential of the chloride ions for the typical concentration differences observed. These chloride ions will therefore be in a (near) equilibrium where it will hold that $J_{\text{Cl}} \sim 0$. Using values from Table 2.1, the Nernst potential for potassium, $V_{\text{Nernst,K}}$, is slightly lower, and equals approximately -93 mV, while the Nernst potential for sodium, $V_{\text{Nernst,Na}}$, is approximately +63 mV. These Nernst potentials are often denoted as E_{K} and E_{Na}, respectively.

Assuming that the membrane has a constant thickness with uniform properties, and assuming that the electrical field inside the cell membrane is homogeneous, it holds for the membrane potential at rest (i.e. the total ionic current is zero)

$$V_{\text{GHK}} = \frac{kT}{q} \ln \frac{P_K[\text{K}^+]_o + P_{Na}[\text{Na}^+]_o + P_{Cl}[\text{Cl}^-]_i}{P_K[\text{K}^+]_i + P_{Na}[\text{Na}^+]_i + P_{Cl}[\text{Cl}^-]_o} \tag{2.10}$$

where P is the permeability coefficient of an ion, with units ms^{-1}, given by

$$P = \frac{D\beta}{d} \tag{2.11}$$

with D the diffusion coefficient, assumed constant, d the membrane thickness, and β the partition coefficient, a measure for the solubility of a substance in lipids. This coefficient describes the ratio of the concentrations of ions from the intra- and extracellular fluid to just inside the membrane. The indices i and o refer to intracellular and extracellular, respectively. Note the reversal of the indices of chloride due to the negative charge carried by this ion. This equation is known as the ("constant field") Goldman-Hodgkin-Katz (GHK) current equation, and was firstly formulated by Goldman (1943) and later modified by Hodgkin and Katz (1949). For a derivation of this equation, we refer to Plonsey [5].

Example

Using the GHK-equation, we will calculate the rest membrane potential of a skeletal muscle fiber of the frog. Typical values for the concentrations of the different ions, in mmol/l, are shown in Table 2.3. Using Eq. 2.10 we find

Table 2.3 Typical values of the primary ion constituents in the intra- and extracellular fluid with the permeabilities at T=20°C. Data refer to skeletal muscle of the frog.

Ion	intracellular	extracellular	permeability coefficient
Na^+	12 mmol/l	145 mmol/l	2×10^{-8} (cm/s)
K^+	155 mmol/l	4 mmol/l	2×10^{-6} (cm/s)
Cl^-	4 mmol/l	120 mmol/l	4×10^{-6} (cm/s)

$$\begin{aligned} V_{GHK} &= \frac{kT}{q} \ln\left[\frac{4P_k + 145P_{Na} + 4P_{Cl}}{155P_k + 12P_{Na} + 120P_{Cl}}\right] \\ &= \frac{kT}{q} \ln\left[\frac{26{,}9 \times 10^{-6}}{790{,}24 \times 10^{-6}}\right] = -85.3 \text{ mV}. \end{aligned}$$

The special structures in the cell membrane through which particular ions can move are known as *ion channels.* The sodium channel is highly selective for sodium, while the potassium channel is highly selective for potassium. We will see later, that these channels can be controlled by the transmembrane voltage, V, to allow the generation of action potentials. Let us first, however, discuss a few more issues related to the situation where the transmembrane voltage is at rest. As we discussed above, the transmembrane voltage, V, as quantified by the GHK equation is typically not equal to either the Nernst potential of sodium or potassium. Therefore, "leaky currents" from potassium

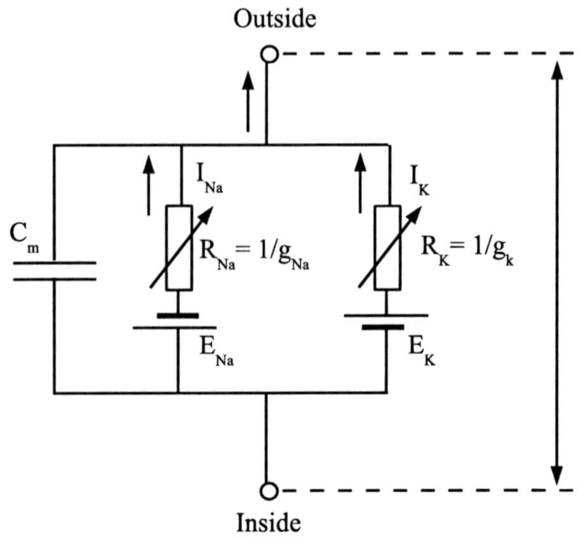

Fig. 2.9 Electrical equivalent circuit of membrane with potassium and sodium channels. The capacitance of the lipid bilayer of the cell membrane is modeled with the capacitance C. The variable conductance of the potassium and sodium channels is indicated with the arrow symbol.

ions leaving the cell and sodium ions entering the cell will be present. These currents are a function of the difference between the Nernst potentials of each ion species and the transmembrane potential, V, and the conductances g_K and g_{Na} for potassium and sodium, respectively. The conductance is the reciprocal of the resistance, with unit (mS cm^{-2}). An electrical equivalent circuit of the membrane with the sodium and potassium conductances is shown in Figure 2.9. A capacitor C_m is used to model the lipid bilayer, and the (variable) resistors $\mathrm{R_{Na}}$ and $\mathrm{R_K}$ model the ion channel conductivities of sodium and potassium. The Nernst potentials for sodium and potassium are modeled with the voltage sources, $\mathrm{E_{Na}}$ and $\mathrm{E_K}$, respectively.

Applying Ohms' law, we now obtain

$$I_{\mathrm{K}} = g_{\mathrm{K}}(V - E_{\mathrm{K}}) \tag{2.12}$$

and

$$I_{\mathrm{Na}} = g_{\mathrm{Na}}(V - E_{\mathrm{Na}}) \tag{2.13}$$

for the leaky currents from the potassium and sodium ions, respectively. These current flows, therefore, are determined by the difference of the transmembrane potential, V and the Nernst potential of potassium, E_K and sodium, E_{Na}, respectively. If this process were allowed to continue forever, the ion concentration gradients across the cell membrane would eventually become zero.

It is the sodium-potassium "pump" (Na-K-pump), localized in the membrane, that compensates for these leaky currents by an active transport mechanism. The Na-K pump transports sodium ions from the inside to the outside, exchanging it with potassium ions, entering the cell. Note, that this pump pumps ions against their concentration gradient. Although not explicitly indicated in the equations for the leaky currents, Eq. 2.12 and Eq. 2.13, the conductances, g, are not a constant, but a rather complicated function of voltage and time. These properties allow the cell to change its transmembrane voltage, generating action potentials. This will be discussed next.

2.4 Action Potential Generation

As follows from the Goldman-Hodgkin-Katz equation 2.10, the main factors that determine the value of the cell membrane potential are the concentration gradients of the several ion species and their actual permeabilities. Excitable cells, nerve cells in particular, have the property that they are able to change very quickly the permeability for certain ion species after application of an adequate stimulus. This leads to a *depolarization* of the cell membrane, and the generation of an *action potential*. The origin of the action potential is in the voltage and time dependent properties of the ion channels for mainly sodium and potassium ions.

During depolarization resulting in action potential generation, the cell membrane voltage changes sign, and the extracellular part becomes negatively charged relative to the intra-cellular fluid. An example of an adequate stimulus for depolarization for a nerve sensitive for touch is deformation of the skin. Hereafter, an action potential will travel, with constant velocity, along the axon. This action potential has a constant amplitude, which is independent of the stimulus strength. This phenomenon is known as the "all-or-nothing" principle. However, the stimulus strength is "translated" into the frequency of the action potentials. Excitable cells, therefore, encode information spatially by differences in the location of the several receptors, and temporally by the frequency of the action potentials. In this process, there is a critical role for some important feedback loops.

The process can be described as follows. After application of an adequate stimulus, the permeability of the cell membrane for sodium ions increases and, if the strength of the stimulus is sufficient, the threshold is reached. This causes the permeability for sodium ions to increase quickly further, to a value many orders of magnitude larger than the permeabilities of the other ion species, especially potassium[13]. Due to this process, the cell membrane potential will reach a value close to the Nernst potential of sodium, which is about + 60 mV.

[13] The permeability for sodium is temporarily regulated by a positive feedback loop, in which depolarization increases the permeability for sodium, which increases the influx of sodium ions, which in its turn further depolarizes the cell membrane, etc.

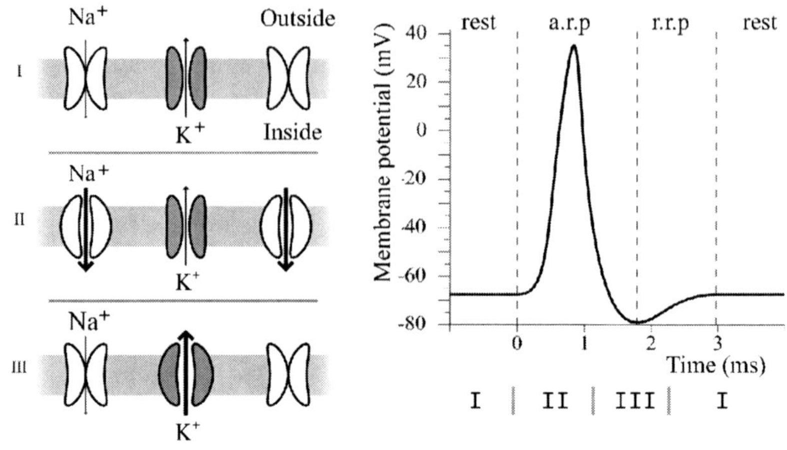

Fig. 2.10 Left: Illustration of the generation of the action potential, with the concurrent changes in the permeability of sodium and potassium. Right: corresponding change in transmembrane voltage. In rest, the potassium conductance is about a hundred times larger than the sodium conductance, which explains why the cell in rest is near the Nernst potential of potassium (I). At the onset of the action potential, the sodium permeability strongly increases (II). There is an undershoot in the membrane voltage during repolarization, due to a temporary increase in potassium conductance (III). Immediately following the beginning of the action potential, the cell membrane is in the absolute refractory period (a.r.p.), during which the cell is insensible for any stimulus. This is followed by the relative refractory period (r.r.p.), during which a new action potential can be generated; only, however, with a stimulus strength much larger than the threshold during the rest membrane potential.

After about 0.3-1 ms, however, the time dependency of the sodium permeability starts to play a significant role, leading to a decrease in the value of P_{Na}. This is followed by an increase in the permeability for potassium ions, which speeds the *repolarization* of the cell membrane, causing a value temporarily lower than the rest cell membrane potential, known as *hyperpolarization*, quickly followed by a return to the rest membrane potential. This cycle is illustrated in Figure 2.10.

The duration of the action potential may vary in different cell types. Furthermore, within a particular cell type we also observe differences in duration. In mammals, the duration of action potentials varies from < 1 ms to 2-3 ms. During generation of the action potentials, cells cannot generate a new action potential. This period is known as the *absolute refractory period.* In most neurons, the absolute refractory period is about 2 ms after the initiation of the action potential. In this period, stimulation will never yield another action potential[14]. Hereafter, there exists a short *relative refractory period*, with a duration up to 5 ms. Due to the *absolute* refractory period, the maximum number of action potential that can be generated per second (the firing rate) is limited to about 500 Hz. Some neurons possess a slightly shorter absolute refractory period, and have fire frequencies up to 1 kHz. This typically occurs only in short durations: bursts[15].

In sum, the action potential arises by a sudden change in the sodium permeability of the cell membrane, caused by depolarization above the threshold. Due to the depolarization, the structure of the channel changes, and the permeability changes. Time is needed to change back to the original structure, Besides these voltage gated channels, various channels exist that are sensitive to particular neurotransmitters (ligand gated channels). These neurotransmitters include dopamine, serotonin, acetylcholine, GABA, AMPA, and many others, as we will discuss in later chapters. As a final remark, if many action potentials are generated, the concentration differences of the ions across the membrane will gradually decrease. Here again the Na-K pump is essential: this pump restores the original concentration differences of the sodium and potassium ions. It can be calculated, though, that the amount of charge transported in the generation of a single action potential is very small. Therefore, thousands of action potentials can be generated before a significant change in the intra- or extracellular ion concentration occurs. The student is advised here to look at Exercise 2.15 that quantifies these concentration differences.

2.4.1 Hodgkin-Huxley Model of the Action Potential

Major contributions to the understanding of the time-course and the generation of the action potential were made by Hodgkin and Huxley. In 1952, Hodgkin and Huxley published five papers that described a series of experiments and an empirical model of an action potential in a squid giant axon [6]. Various elements made the squid giant axon an ideal candidate for studying the properties of membrane channels. The squid axon can survive a long time in sea water and it has only two types of voltage-time-dependent permeable channels. The large size of the squid axon (diameter as large as 1000 μm) makes it possible to insert an axial electrode directly into the axon. By placing a second electrode outside the axon, it is possible to measure the transmembrane voltage. This setup also allows stimulation of the axon by injecting a current with extracellular electrodes. Various additional experiments are feasible, as well. For instance, one can replace the extracellular fluid with a fluid with a different ionic composition. Sodium currents can be blocked with tetrodotoxin (TTX). The major revolution, however, was

[14] This is caused by the time needed for an ion channel to return to its baseline state.

[15] Auditive evoked potentials, that can be realized using a click-stimulus, can induce these bursts in the spiral ganglion, at the origin of the acoustic nerve (N VIII).

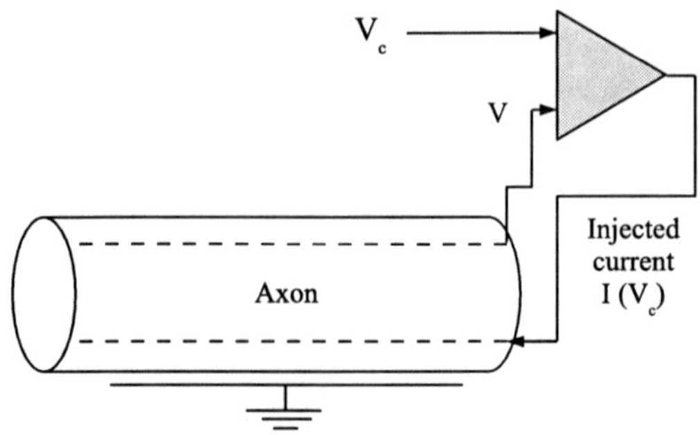

Fig. 2.11 Voltage-clamp experiment on the axon. The injected current is proportional to the difference $V_c - V$, and keeps the membrane potential at V_c. By analysis of the instantaneous changes in the current as a function of steps in V_c, Hodgkin and Huxley were able to describe the various voltage and time dependent changes in the ionic conductances.

the use of the voltage-clamp technique. Voltage-clamping is the technique by which a current is injected into the axon so that a predetermined fixed voltage is maintained across the membrane. This injected current is the mirror image of the current generated by the cell membrane at that potential. This is illustrated in Figure 2.11. In this way, by setting the membrane potential to different values, and using specific blockers and ion substitution protocols, information is obtained about the voltage dependent characteristics of the transmembrane currents.

Hodgkin and Huxley (and Katz, who co-authored one paper, and was a co-worker in several experiments) studied the changes in membrane resistance as a function of time and applied transmembrane voltage. Using the constant voltage differences across the membrane (voltage clamp conditions), they were able to measure changes in the transmembrane currents as a function of changes in membrane resistance. These membrane resistance changes were caused by the opening and closing of the voltage dependent sodium and potassium channels. An example of such an experiment is shown in Figure 2.12.

By analyzing the estimated sodium and potassium conductances (g_{Na} and g_K, respectively) from voltage clamp pulses of different amplitudes and durations, combined with the use of blockers and variations in the extracellular ion composition, Hodgkin and Huxley were able to obtain a complete set of nonlinear empirical equations that described the time-course of the action potential generation. The empirical equation used to model the sodium and potassium conductances is a system of four coupled nonlinear ordinary differential equations. These equations describe the changes in membrane voltage as function of the voltage dependent membrane conductances. For instance, the potassium current, I_k, is described as

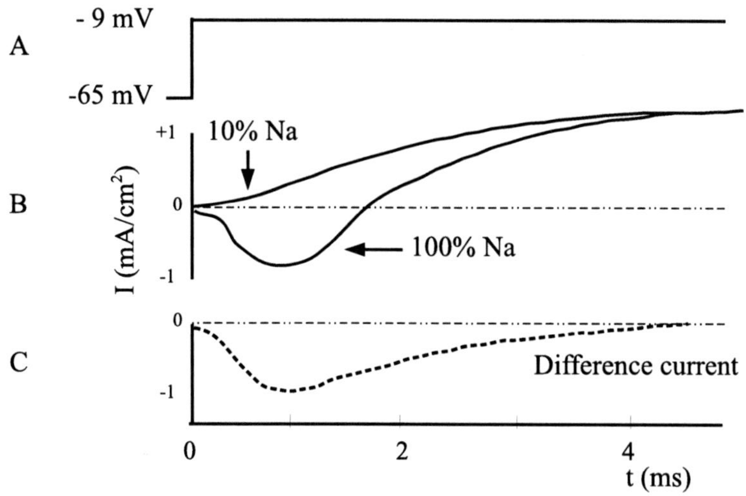

Fig. 2.12 Clamp current in response to a step from -65 mV to -9 mV. In A, the step voltage is indicated. This experiment involved replacing the extracellular sodium (100%) with a 10% solution (B). The difference current, C, is now essentially the sodium current. Illustration adapted from Hodgkin [7] and Figure 4.11 in [8].

$$I_{\mathrm{K}} = g_K(V - E_{\mathrm{K}}) = \overline{g}_K n^4(V - E_{\mathrm{K}}) \tag{2.14}$$

with g_K the conductance for the potassium ions. Compare this with Eq. 2.12, where we have now made the potassium conductance g_K a function of the maximum potassium conductance, $\overline{g}_K$ and the gating variable, n. The rate of change of the gating variable is given by

$$\frac{dn}{dt} = \alpha_n(V)(1-n) - \beta_n(V)n \tag{2.15}$$

with α_n and β_n exponential functions of the membrane voltage, V. This means that n is a time dependent function of the membrane voltage, where n defines the time course of the opening and closing of the potassium channel. A similar expression was found for the kinetics of the sodium channel. There is an important qualitative difference between the sodium and potassium conductance changes, however. In response to a sustained voltage, the change in sodium conductance is *transient*, and only lasts a few milliseconds, whereas the changes in the potassium conductance is sustained. The expression that describes the sodium conductance, therefore, has *two* gating variables: one that describes the opening of the sodium channel (the activation variable), m, and one that describes the closing (or inactivation) of the sodium channel, h. Hodgkin and Huxley found empirically (by fitting the data obtained with the voltage clamp experiments) that they had to raise m to the power of three, to arrive at

$$I_{\mathrm{Na}} = g_{Na}(V - E_{\mathrm{Na}}) = \overline{g}_{Na} m^3 h(E - E_{\mathrm{Na}}) \tag{2.16}$$

where m and h are defined as

$$\frac{dm}{dt} = \alpha_m(V)(1-m) - \beta_m(V)m, \tag{2.17}$$

$$\frac{dh}{dt} = \alpha_h(V)(1-m) - \beta_h(V)h. \tag{2.18}$$

Hodgkin and Huxley also succeeded in finding explicit expressions for α and β. The complete set of space-clamped Hodgkin-Huxley equations is now

$$C\frac{dV}{dt} = I - \overline{g}_k n^4(V - E_k) - \overline{g}_{Na} m^3 h(V - E_{Na}) - g_L(V - E_L) \tag{2.19}$$

with n, m, h as defined previously, g_L the leak conductance, and

$$\alpha_n(V) = 0.01\frac{10-V}{\exp(\frac{10-V}{10}) - 1}, \tag{2.20}$$

$$\beta_n(V) = 0.125\exp(\frac{-V}{80}), \tag{2.21}$$

$$\alpha_m(V) = 0.1\frac{25-V}{\exp(\frac{25-V}{10}) - 1}, \tag{2.22}$$

$$\beta_m(V) = 4\exp(\frac{-V}{18}), \tag{2.23}$$

$$\alpha_m(V) = 0.07\exp(\frac{-V}{20}), \tag{2.24}$$

$$\beta_m(V) = \frac{1}{\exp(\frac{30-V}{10}) + 1}. \tag{2.25}$$

The formulation of these equations was a major breakthrough in the understanding of the dynamics of the action potential generation[16]. It is interesting to note, that at the time of the formulation of the Hodgkin-Huxley equations, ion channels had not been identified as such. It was only recently, that the processes involved in the selectivity of ion channels were elucidated. Major contributions were made by Roderick MacKinnon, who studied a potassium selective channel in bacteria, and Peter Agre, who discovered aquaporines. These aquaporines are water-channel proteins that move water molecules through the cell membrane. Both were awarded the Nobel Prize in Chemistry in 2003 "for discoveries concerning channels in cell membranes". A

[16] The notation used here is historical. In the standard form (as displayed in most textbooks), the differential equations for the activation variables n, m, and h are typically written as $\dot{n} = (n_\infty(V) - n)/\tau_n(V)$ with $n_\infty = \alpha_n/(\alpha_n + \beta_n)$, where $\tau_n = 1/(\alpha_n + \beta_n)$. Expressions are similar for m and h.

very good text for a deeper understanding of the Hodgkin-Huxley equations and the resulting dynamical behavior of neurons and neuronal networks is Izhikevich [9].

2.5 Channelopathies

Various diseases exist, in which the ion channels are malfunctioning. These diseases are collectively known as channelopathies [10, 11]. These channelopathies are ion channel defects due to a genetic mutation in the ion channel genes[17]. Over 340 human genes are known to code for ion channels. Channelopathies have been identified in both voltage gated channels (Na^+, K^+, Ca^{2+}, Cl^-) and various ligand gated channels. Ligand gated channels are ion channels that are controlled by a ligand: a particular neurotransmitter (for instance, dopamine, acetylcholine, serotonin, GABA)[18]. The kind of change that occurs in the affected ion channel is variable, and different mutations can result in distinct functional defects. Most of these channelopathies are caused by a single mutation, responsible for a change in the protein structure in the channels. In general, all channelopathies cause changes in the membrane excitability of cells, whether in the central or peripheral nervous system, or the skeletal muscles. In most cases, channelopathies only affect a single organ system, since the ion channel expression is often tissue specific.

For instance, in various epilepsy syndromes, mutations in channels have been found. As an example, mutations in the gene SCN1B, which encodes a voltage-gated sodium-channel subunit, are associated with generalized epilepsy with febrile seizures plus. The apparent effect of these mutations is to allow passage of an increased sodium current, which would lead to a greater depolarization during the action potential and an increased tendency to fire repetitive bursts. Mutations in KCNQ2 and KCNQ3, which both encode potassium channels, are associated with benign familial neonatal convulsions. Familial hypokalaemic periodic paralysis is characterized by intermittent muscle weakness, due to abnormal functioning of a sodium or calcium ion channel. In some forms of migraine, the familial hemiplegic migraine, an abnormal Ca-channel has been identified. Currently, mutations in more than 60 ion genes are known to be responsible for channelopathies. An overview of some common channelopathies is shown in Table 2.4.

---

[17] For completeness, we mention that particular autoimmune diseases may also cause abnormal ion channel functioning. An example is myasthenia gravis (to be discussed in more detail later), in which antibodies are found directed against the acetylcholine receptor, which is a ligand gated ion channel. In the literature, some authors share these autoimmune induced abnormalities in ion channel dysfunction to the channelopathies, as well.

[18] Ligand gated channels are named after the controlling ligand, and typically referred to as a *receptor*. Therefore, we have an acetylcholine receptor or a GABA-receptor. Voltage gated channels are typically named according to the ion that is transported through the channel: sodium channels, chloride channels, calcium channels.

Table 2.4 Overview of some channelopathies. This is only a very short list, since at present more than 50 channelopathies have been identified. HypoPP= familial hypokalaemic periodic paralysis. AD=autosomal dominant.

	channel involved
CENTRAL NERVOUS SYSTEM	
AD myoclonic epilepsy	Cl^-
Benign familial neonatal seizures	K^+
Episodic ataxia type 1	K^+
Episodic ataxia type 2	Ca^+
Benign neonatal-infantile seizures	Na^+
Familial hemiplegic migraine	Ca^+
Spinocerebellar ataxia type 6	Ca^+
SKELETAL MUSCLE AND NERVE	
Myotonia congenita	Cl^-
Paramyotonia	Na^+
HypoPP	Ca^+ or Na^+
periodic paralysis	Na^+
Familial generalized myokymia	K^+
EXOCRINE GLANDS	
cystic fibrosis	Cl^-
HEART	
short QT syndrome	K^+
Long QT syndrome	Na^+, K^+, Ca^+ (depending on type LQT (1-10))

2.6 Electrolyte Abnormalities

Since the resting potential of excitable cells, in particular neurons, and, as we will discuss in more detail in a later chapter, muscle cells, is mainly determined by the Nernst potential of potassium, changes in the intra- and or extracellular potassium concentration will directly affect the excitability properties. Similar arguments hold for changes in the sodium concentration, since this ion is highly relevant in the generation of the action potential. In clinical practice, these situations are very often encountered.

In hyponatraemia, for instance, the amplitude of the action potentials is severely reduced, which can even result in the inability for cells to generate action potentials or transport them sufficiently. Clinically, this can result in loss of reflexes, convulsions and coma. Although one can wonder why convulsions can occur (typically due to increased synchronization and increased spiking of neuronal populations), we should take into account here that neural assemblies are comprised of cells with different physiological characteristics. Therefore, in hyponatraemia, it is not unlikely that particular inhibitory (modulating) neurons reduce their activity before other cell populations (e.g. pyramidal cells in the motor cortex) do. Therefore, an imbalance between neuronal populations could result in abnormal increased activity, that may clinically manifest as seizures. Another contributing factor, that is a result of shifts in ion concentrations, is osmosis. Severe hyponatraemia may cause a shift of water from the plasma into the brain cells. Causes of hyponatraemia are varied, including water intoxication, adrenal insufficiency and abnormal (increased) secretion of antidiuretic hormone (ADH). Water intoxication is not uncommon in long-duration runners. In the Boston Marathon, Almond et al. [12] found a hyponatraemia (135 mmol/liter or less) in about 13% of the 488 runners at the finish line. About 0.6% had critical hyponatraemia, with levels of 120 mmol per liter or less, which is life threatening.

In hypernatraemia, clinical manifestations can, in mild cases, be subtle, consisting of e.g. weakness or edema. Severe hypernatraemia, may also result in coma and seizures. Causes include diminished water intake, the use of diuretics or lithium. Another, less common cause, is the inability of the kidneys to retain water, due to diminished production of anti-diuretic hormone from the hypophysis or if the kidneys do not react to this hormone. This condition is known as diabetes insipidus.

Hyper- and hypokalaemia are conditions also often encountered in clinical practice. Increased serum potassium can be caused by abnormal intake, while a common cause of low serum potassium is reduced intake or the use of particular diuretics. Abnormal secretion of aldosterone will also result in abnormal serum potassium. Clinical symptoms are varied, and include cardiac rhythm disturbances and muscle weakness. Here, again, there is overlap in clinical symptomatology; therefore, it is not always easy to differentiate between hyper- and hypokalaemia based on the symptoms and signs observed.

It is not straightforward to conceptualize all the signs and symptoms in electrolyte disturbances based on the previous discussion of the origin of the membrane potential and the generation of action potentials. A non-trivial mechanism, for instance, is that particular sodium channels can inactivate in hyperkalaemia. Therefore, although one would in first instance expect an increase in excitability, hyperkalaemia may also result in reduced excitability of cells, in particular muscle tissue.

2.7 Summary

In this chapter the origin of the membrane potential and the generation of action potentials was presented. You understand the Nernst and Goldman-Hodgkin-Katz equation and the generation of the Nernst potential. You are familiar with concepts as electrostatic and diffuse forces. You are also familiar with the the processes that occur in the generation of an action potential, including concepts as the absolute refractory period and relative refractory period. You understand the basics of the Hodgkin-Huxley equations. You have some understanding of channelopathies. We also shortly discussed the clinical effects of increased and decreased potassium and sodium concentrations. For students who are interested in additional information, see e.g. [13, 14, 15].

Problems

2.1. Can you mention some examples of input stimuli that are typically processed consciously or unconsciously?

2.2. Present some examples of feedback in the nervous system.

2.3. Try to derive the Nernst equation yourself. What is meant with the expression: "*The ion has two masters*"?

2.4. Which ions play the major role in the cell membrane potential? In animal experiments the animal can be killed painlessly by intravenous injection of a high dose of KCl. Can you explain this?

2.5. A red blood cell has the ion distribution as indicated in Table 2.5. Calculate the rest membrane potential at 37°C.

Table 2.5 Ion distribution in a red blood cell.

ion	intracellular (mmol/l)	extracellular (mmol/l)	activity (cm/s)
Na^+	19	155	2×10^{-8}
K^+	136	5	2×10^{-6}
Cl^-	78	112	3.9×10^{-6}

2.6. Describe the permeability changes and ionic fluxes which occur during the action potential (AP). What factors affect the timing of ion conductance during the AP (e.g., What opens an ion gate or closes it)?

2.7. Following an action potential spike, the neuron briefly becomes more hyperpolarized (that is, more negative than -70 mV). What is the electrochemical basis for this hyperpolarization? Which process returns the membrane potential to -70 mV?

2.8. Some references cite that the intracellular Na^+ concentration in a resting neuron is about 5 mmol/l, rather than 15 mmol/l. Calculate the effect of lowering $[Na^+]_i$ to 5 mmol/l. Does this difference have a major effect on resting membrane potential?

2.9. What happens to the resting membrane potential if you lower the concentration of K^+ on the inside of the cell from 150 mmol/l to 135 mmol/l? (Keep the other factors constant).

2.10. In what direction do the electrical and the diffuse forces work on the potassium ions during the phase of the action potential when the membrane is depolarized (say, at + 40 mV)?

2.11. The transmembrane potential
a. is temperature dependent
b. is about 70 mV (absolute value)
c. is mainly determined by the concentration gradient of potassium ions
d. All of the above is true.

2.12. The Na-K-pump:
a. realizes the repolarization phase of the action potential
b. maintains the membrane rest potential
c. is a function of the electrical characteristics of the cell membrane.
d. None of the above.

2.13. The action potential arises primarily because
a. the Na^+ -permeability increases
b. the K^+ -efflux increases
c. the Cl^- -efflux increases.
d. All of the above are true

2.14. The period that an axon membrane cannot be stimulated is known as
a. the relative refractory period
b. the relative insensitive period
c. the absolute refractory period
d. the absolute insensitive period

2.15. Consider a piece of an axon with length $l = 10$ cm and radius r=10 µm. We model the membrane as a cylindrical parallel-plate capacitor with thickness $d = 10^{-8}$ m. The capacity C is given by $C = \mathrm{k}\epsilon_0\ \mathrm{A/d}$. Set the relative permittivity of the dielectric material k=3. $\epsilon_0 = 8.85 \times 10^{-12}\ \mathrm{C^2/N\ m^2}$.

a. Calculate the surface and capacity of this axon with length l and radius r.
b. If a single action potential is generated, then the membrane potential changes from the resting potential -70 mV to + 30 mV. How much charge is transported during the generation of this single action potential? (Remember: Q = CV)
c. What is the change in concentration of the sodium ions *in* the cell due to the generation of this action potential? Use for the baseline intracellular sodium concentration 15 mmol/l.

References

1. Catani, M., Ffytche, D.H.: The rises and falls of disconnection syndromes. Brain 128, 224–239 (2005)
2. Ramachandran, V., Blakeslee: Phantoms in the Brain. HarperCollins (2003)
3. Rizzolatti, G.: The mirror neuron system and its function in humans. Anat Embryol (Berl) 210(5-6), 419–421 (2005)
4. Churchland, P.S., Sejnowski, T.J.: The computational brain. MIT Press, Cambridge (1999)
5. Plonsey, R., Barr, R.C.: Bioelectricity: a quantitative approach. Springer, Berlin (2007)
6. Hodgkin, A.L., Huxley, A.F.: A quantitative description of membrane current and its application to conduction and excitation in nerve. J. Physiol (Lond) 117, 500–544 (1952)
7. Hodgkin, A.L.: The Croonian lecture: Ionic movements and electrical activity in giant nerve fibers. Proc. Roy. Soc. London, Series B 148(930), 1–37 (1958)
8. Beuter, A., Glass, L., Mackey, M.C., Titcombe, M.S. (eds.): Nonlinear Dynamics in Physiology and Medicine. Springer, Heidelberg (2003)
9. Izhikevich, E.M.: Dynamical Systems in Neuroscience. MIT Press, Cambridge (2007)
10. Ashcroft, F.M.: Ion Channels and Disease. Academic Press, London (2000)
11. Goodman, B.E.: Channels active in the excitability of nerves and skeletal muscles across the neuromuscular junction: basic function and pathophysiology. Adv. Physiol. Educ. 32(2), 127–135 (2008)
12. Almond, C.S.D., Shin, A.Y., Fortescue, E.B., Mannix, R.C., Wypij, D., Binstadt, B.A., Duncan, C.N., Olson, D.P., Salerno, A.E., Newburger, J.W., Greenes, D.S.: Hyponatremia among runners in the boston marathon. N Engl. J. Med. 352(15), 1550–1556 (2005)
13. Bower, J.M., Beeman, D.: The book of Genesis. Exploring realistic models with the general neural simulation system (2003), `http://www.genesis-sim.org//GENESIS`
14. Dayan, P., Abbott, L.F.: Theoretical Neuroscience. Computational and Mathematical Modeling of Neural Systems. MIT Press, Cambridge (2001)
15. Enderle, J.D., Blanchard, S.M., Bronzino, J.D.: Introduction to Biomedical Engineering. Elsevier Academic Press, Amsterdam (2005)

Chapter 3
Propagation of the Action Potential

3.1 Introduction

Besides the generation of action potentials, nervous tissue is especially designed for the *propagation* of these action potentials. In fact, this is one of the prime functions of the central and peripheral nervous system: information transport.

We discussed in the former chapter how the action potential is generated. Due to this local change in voltage, currents will flow, partially intracellular and partially extracellular. These currents will subsequently change the transmembrane voltage of the neighboring parts of the cell membrane and, if threshold is reached, action potentials will be generated in these neighboring parts, due to the action of the voltage gated channels. In this chapter, we will look into this process in more detail.

Action potentials are conducted by both passive and active processes. Passive conduction of the action potential requires no energy, and is realized by those parts of the axonal membrane that are in between the voltage dependent channels. If an action potential is generated at a particular position at the axon, the voltage change that is created will induce currents, that are not only conducted inside the axonal cytoplasm, but also dissipated by leakage out across the axon membrane. The conduction allowed by these passive membrane properties is known as *electrotonic conduction*. However, action potential propagation cannot rely on electrotonic conduction, only. To "travel" the actual distance (sometimes up to one meter) along the axons and dendrites, active processes are needed, as well. This is realized in *regenerating axons*, allowing the creation of new action potentials. Both electrotonic conduction and action potential regeneration are essential for proper information transport in the central and peripheral nervous system and Nature has designed various mechanisms to optimize it. Here, we will start with an extensive discussion on the passive membrane properties.

M.J.A.M. van Putten: Essentials of Neurophysiology, BIOMED, pp. 33–52.
springerlink.com

3.2 Passive Membrane Properties

Let us consider[1] a very long cylindrical segment of a neuron, a part of which is shown in Figure 3.1. The membrane potential, $V(x,t)$, along this segment is determined by solving a partial differential equation, the cable equation, that describes how the currents entering, leaving and flowing within a neuron affect the rate of change of the membrane potential.

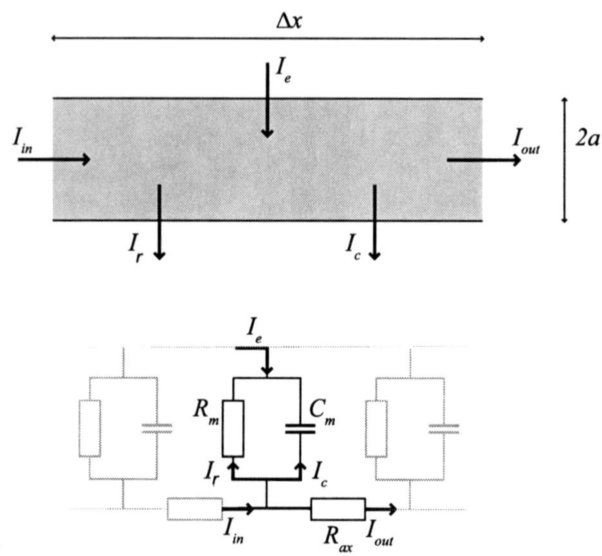

Fig. 3.1 Top: Segment of a neuron used in the derivation of the cable equation. The longitudinal, total membrane current $I_m = I_r + I_c$, and electrode current, I_e, that determine the rate of change of the membrane potential within this segment, are denoted. The segment has length Δx and radius a. Bottom: Electrical model of an axon segment. The total membrane resistance is modeled as R_m, with membrane capacitance C_m. The axial resistance is R_{ax}.

At the left of this segment with length Δx, current enters the axon. At the right, current leaves the segment. The (displacement) current that charges the membrane due to its capacitative properties is indicated by I_c. In addition, two transmembrane currents are shown. The electrode current, I_e, is a current that is provided by a possible external stimulus (for instance to stimulate the axon). The resistive membrane current, I_r is the current that may "spontaneously" leave the segment, due to its finite resistive properties. Applying Kirchoff's law to the currents shown in Figure 3.1, we obtain

[1] This section 3.2, treating passive membrane properties, is strongly inspired by the treatment of the cable equation as presented in Chapter 6 in Dayan and Abbott, Theoretical Neuroscience, the MIT Press [1].

$$I_c = I_{in} - I_{out} + I_e - I_r. \tag{3.1}$$

We will now write the various currents as a function of membrane voltage, axon segment dimensions, and physiological properties as axial resistance, membrane resistance and membrane capacitance. Before we actually do so, however, we introduce *specific units*: parameters that are independent of the cell dimensions. Since the axial resistance of a cylindrical compartment is proportional to its length and inversely proportional to its cross-sectional area, we define r_{ax} as the *specific axial resistance*, with units Ω·cm. For a cylindrical compartment, the membrane resistance is inversely proportional to the area of the cylinder, so we define a *specific membrane resistance* r_m with units Ω·m^2. Finally, membrane capacitance is proportional to the area, and we therefore define the *specific membrane capacitance* c_m with units µF/cm^2. The total membrane capacitance C_m is then given by the product of the specific membrane capacitance and the area of the cylinder: $C_m = 2\pi a \Delta x c_m$.

Using Ohm's law, we can now write for the current I_{in} that enters the segment (with radius a)

$$I_{in} = -\frac{\pi a^2}{r_{ax}} \frac{\partial V}{\partial x} \tag{3.2}$$

and a similar expression for the current that leaves the segment[2].

Further writing for the resistive membrane current, $I_r = i_r 2\pi a \Delta x$ and for the electrode current $I_e = i_e 2\pi a \Delta x$, Eq. 3.1 becomes

$$C_m \frac{\partial V}{\partial t} = 2\pi a \Delta x c_m \frac{\partial V}{\partial t} = -\left(\frac{\pi a^2}{r_{ax}} \frac{\partial V}{\partial x}\right)_x + \left(\frac{\pi a^2}{r_{ax}} \frac{\partial V}{\partial x}\right)_{x+\Delta x} - 2\pi a \Delta x (i_r - i_e). \tag{3.3}$$

Dividing both sides of this equation by $2\pi a \Delta x$, assuming that the specific axial resistance r_{ax} is constant, and letting $\Delta x \to 0$, we obtain the cable equation

$$c_m \frac{\partial V}{\partial t} = \frac{1}{2 a r_{ax}} \frac{\partial}{\partial x} \left(a^2 \frac{\partial V}{\partial x}\right) - i_r + i_e. \tag{3.4}$$

We will make a few more assumptions for further simplification. First, we assume that the radii of the cable segments used to model a neuron are constant. This allows us to take the factor a^2 out of the derivative. Second, we set the membrane potential v relative to the resting potential, V_{rest}, $v = V - V_{\text{rest}}$. We now write for the resistive membrane current $i_r = v/r_m$, with r_m the specific membrane resistance. This is a linear approximation of the membrane current, that is valid only if the membrane potential does not depart too much from the resting membrane potential. With these modifications, the cable equation for v is now

$$c_m \frac{\partial v}{\partial t} = \frac{a}{2 r_{ax}} \frac{\partial^2 v}{\partial x^2} - \frac{v}{r_m} + i_e. \tag{3.5}$$

[2] This equation is obtained from $\Delta V = -I_{in} r_{ax} \Delta x / \pi a^2$, and letting $\Delta x \to 0$.

It is convenient to multiply this equation by r_m, and to introduce the membrane time constant

$$\tau_m = r_m \cdot c_m \tag{3.6}$$

and the electrotonic length (also known as the membrane length constant or space constant)

$$\lambda = \sqrt{\frac{a r_m}{2 r_{ax}}}. \tag{3.7}$$

We now obtain the final expression for the cable equation for v, expressed in terms of τ_m and λ:

$$\tau_m \frac{\partial v}{\partial t} = \lambda^2 \frac{\partial^2 v}{\partial x^2} - v + r_m i_e. \tag{3.8}$$

The factor λ has the dimensions of length and determines how far the electrotonic spread of the membrane potential is able to influence neighboring regions. Let us calculate the value of λ for a fiber with radius 0.5 µm, which is a typical value for an unmyelinated fiber (0.1-1 µm), with a specific membrane resistance $r_m = 2 \cdot 10^3$ Ωcm^2 and specific axial resistance $r_{ax} = 100$ Ωcm. This results in $\lambda \sim 0.3$ mm. As we will learn shortly, this implies that the action potential will decay to approximately a third of its value at a distance of 0.3 mm from the point where it was generated.

In order to solve the cable equation, Eq. 3.8, we need to define the appropriate boundary conditions. For instance, what is the membrane potential at the end of the axon? When axons branch, longitudinal currents must be preserved, i.e. the current that enters the point where the axon branches must equal the sum of the currents that leave. Furthermore, discontinuities in voltage are not allowed. For real-world neurons, with their complex morphology and variations in biophysical properties, the cable equation is often solved numerically, using multi-compartment techniques. Realistic models take into account the different lengths and diameters of dendrites, axons and their branches. It is not uncommon to apply more than 100 compartments for a single neuron. Since this is not a book on computational and mathematical modeling, these issues will not further be discussed. It is realized, however, that these techniques not only contribute to an improved understanding of basic neurophysiology, but also to clinical medicine. This includes computational modeling of abnormal circuitry, as present in many central nervous system disorders, for instance Parkinson's disease or epilepsy. Interested students who wish to learn more about theoretical neuroscience, and modeling of neural systems are referred to the literature [1, 2, 3, 4].

3.2.1 Steady-State Solution of the Cable Equation

If we solve the cable equation, the solutions we obtain for the membrane voltage v are functions of both position and time. There are special situations,

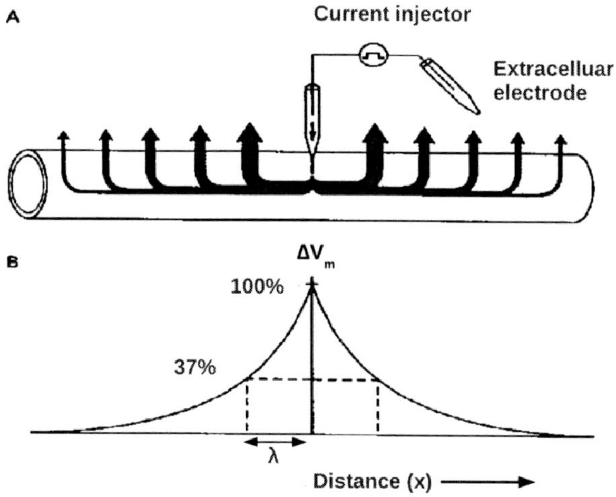

Fig. 3.2 A. Illustration of current injected at $x = 0$. There is an exponential decay of the voltage along the axon, as illustrated in graph B. Adapted from Eric R. Kandel and James H. Schwartz and Thomas M. Jessell, Principles of Neural Science [5], McGraw-Hill, © 2000.

however, that result in steady-state solutions, that are independent of time. Let us, therefore, inject the axon with a constant current. If we now wait sufficiently long, the membrane potential might have settled to a constant value. In this case, therefore, the cable equation will have a time-independent solution and the voltage difference across the membrane is a function of position, only. This gives $\partial v/\partial t = 0$. Then Eq. 3.8 reduces to

$$\lambda^2 \frac{\partial^2 v}{\partial x^2} = v - r_m i_e. \tag{3.9}$$

Furthermore, we will inject a current i_e at one position, $x = 0$, only. Outside the place of the current injection we therefore have $i_e = 0$. This is illustrated in Figure 3.2. Now we will solve Eq. 3.9, assuming an axon that is infinitely long, that results in[3]

[3] The general solution for the steady-state voltage as a function of the distance x, away from the injection site, is given by $V_m(x) = A\exp^{x/\lambda} + B\exp^{-x/\lambda}$ where A and B depend on the boundary conditions. For a finite cylinder, that is sealed at $x = l$, the solution will therefore be different. Here, axial current is prevented from flowing across the boundary at $x = l$ (*Neumann conditions*). Another boundary condition that is often encountered is a finite axon with one of the boundaries clamped to the resting membrane potential (*Dirichlet condition*). In both these conditions, the membrane potential as a function of the distance along the axon is described differently. This, however, will not be further discussed; the interested reader is referred to the literature [1, 3, 4].

$$v(x) = v_o e^{-|x|/\lambda}. \tag{3.10}$$

The membrane potential, therefore, falls of exponentially with distance from the site of the action potential (or current injection in the case of an external stimulus). Note, that inserting $x = \lambda$, we find that the membrane voltage has dropped to 37% of its original value[4] after a distance equal to the length constant. The larger λ, the further down the voltage spreads. It can also be viewed as a measure for preservation of form, since the larger λ, the longer the distance over which the shape of the action potential is relatively preserved (that is why it is also known as space constant). The electrotonic conduction distance, as an effective means for communication, is typically only 0.2-15 mm. This implies that at distances longer than 1-10 mm an amplifier must be present to allow further transmission of the action potential. This latter process is realized by the active membrane properties, discussed in the next section. The space constant λ can also be written as a function of membrane and axial resistance for a piece of dendrite or axon with length l. Using the expressions that relate the membrane resistance R_m to the specific membrane resistance r_m

$$R_m = \frac{r_m}{2\pi \cdot a \cdot l} \tag{3.11}$$

and the expression that relates the specific axial resistance (r_{ax}) to the axial resistance R_{ax}

$$R_{\text{ax}} = \frac{r_{ax} \cdot l}{\pi a^2} \tag{3.12}$$

we obtain

$$\lambda = l\sqrt{\frac{R_m}{R_{ax}}}. \tag{3.13}$$

Note: many textbooks and journal papers define the resistance and capacitance in terms of that for a unit length of cable having a specified diameter, where $R_m = r_m/l$, $C_m = c_m \cdot l$ and $R_a = r_a \cdot l$.

3.2.2 The Membrane Time Constant

Consider again the cable equation, Eq. 3.8, with the membrane time constant $\tau_m = r_m \cdot c_m$. What is the biophysical interpretation of this parameter? Comparing Fig. 3.1, we see that the membrane time constant is defined by the product of the membrane resistance and capacitance. If current flows through this circuit, the rate of change of the voltage across this piece of the cell membrane is a defined by τ_m. Therefore, currents that flow across the membrane gradually change the membrane potential, due to its capacitative characteristics.

---

[4] Students may ask what the original value v_0 is. It can be derived that $v_0 = I_e r_m/(4\pi a\lambda)$, with I_e the current per unit area injected, as discussed in [1].

To quantify this local change in voltage, we will now assume a membrane potential that is spatially uniform. In this case, there is no voltage gradient in the x-direction along the membrane, and $\partial^2 v/\partial x^2 = 0$. Application of a current pulse with duration t and using Eq. 3.8, that is now reduced to

$$\tau_m \frac{\partial v}{\partial t} = -v + r_m i_e \tag{3.14}$$

we obtain

$$v(t) = v_\infty \left(1 - \exp^{-t/\tau}\right) \tag{3.15}$$

with v_∞ the steady-state value of the membrane potential. This expression, therefore, defines how fast the membrane potential changes after the current injection. The value of the membrane time constant, τ is essential here: the larger this value, the slower membrane potentials change in response to an (injected) current. After the current pulse ends, the membrane potential declines exponentially, according to

$$v(t) = v_\infty \exp^{-t/\tau} . \tag{3.16}$$

The values of the membrane resistance and capacitance differ between neurons, and are, in part, a function of the their size. Small nerve cells tend to have longer time constants, and large cells brief time constants. Typical values are in the range from 1-100 milliseconds.

3.2.3 Conduction Velocity

In "real-life", the membrane potentials will be both space and time dependent. Action potentials will therefore result in both spatial and temporal changes in the membrane voltage. If we could travel along with the action potential while it is electrotonically conducted along the axon, we would observe that the action potential widens and becomes lower in amplitude. The velocity with which these two events occur, are defined by the length constant, λ and the membrane time constant, τ. We will now try to find an expression that relates the action potential conduction velocity to these two parameters. A measure of the speed of electrotonic spread can be obtained as follows. Let us apply an instantaneous current pulse at the point $x = 0$ and at the time $t = 0$, that delivers an amount of charge $I_e \tau_m$. This pulse of current models our action potential, generated at $x = 0$. If we now solve Eq. 3.8, we obtain

$$v(x,t) = \frac{I_e R_\lambda}{\sqrt{4\pi t/\tau_m}} \exp\left(-\frac{\tau_m x^2}{4\lambda^2 t}\right) \exp\left(-\frac{t}{\tau_m}\right) \tag{3.17}$$

with $R_\lambda = r_m/2\pi a\lambda$. This equation describes the time and spatial dependence of the membrane potential, v as a result of our current pulse. For the

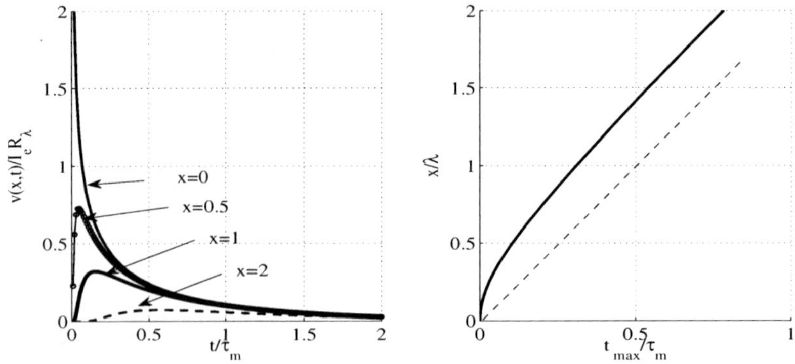

Fig. 3.3 Left: Membrane potential in response to a pulse of current injected at the point $x = 0$ at time $t = 0$. The time dependence, displayed as a function of t/τ_m, is calculated for different measurement sites (illustrated for x=0.5, x=1 and x=2). The farther away the measurement site is located, the later the maximum value is reached. Right: The solid line shows the relationship between a particular measurement position (x/λ) and the time ($t_{\max}/\tau_m$) that the potential has its maximum value at this location. Note that for relative large values of the distance x/λ the velocity is constant. The dashed line corresponds to the velocity $2\lambda/\tau_m$. Values used in the simulation are $r_m = 1$ MΩ mm^2, specific membrane capacitance $c_m = 10$ nF/mm^2, r_{ax}=1 kΩmm and a radius $a = 2\ 10^{-3}$ mm. This yields $\lambda =$ 1 mm, $R_\lambda \sim 8 \cdot 10^7$ Ω and $\tau_m = 10$ ms. Illustration based on Fig. 6.8 in [1].

derivation, we refer to the literature (Tuckwell, 1988). By computing the time $t_{\max}$ when the maximum potential occurs at a given spatial location, we obtain a measure for the velocity of the spread of electrotonic disturbances, as illustrated in Figure 3.3. This is done by setting the time derivative of $v(x, t)$ in Eq. 3.17 to 0. This yields

$$t_{\max} = \frac{\tau_m}{4}\left(\sqrt{1 + 4(x/\lambda)^2} - 1\right). \tag{3.18}$$

If we now consider a solution for large x, it is straightforwardly found that $t_{\max} \sim x\tau_m/2\lambda$. This implies, that the velocity of the action potential is approximated by

$$v_{\text{ap}} \sim 2\frac{\lambda}{\tau_m}. \tag{3.19}$$

Using realistic values of $\lambda = 1$ mm and $\tau_m = 0.01$ s, we find $v \sim 0.2$ m/s. Note, however, that for small values of x the velocity is faster. Therefore, the propagation velocity for an action potential along an unmyelinated, non-regenerating, axon is proportional to the ratio of the electrotonic length constant to the membrane time constant. Using Eq. 3.7 and $\tau_m = r_m c_m$, we can also write

$$v_{ap} \sim 2\frac{\lambda}{\tau_m} = \sqrt{\frac{2a}{c_m^2 r_{ax} r_m}} \tag{3.20}$$

Therefore, the velocity is proportional to the square root of the radius, a. Higher speeds, therefore, are realized in thick axons. As discussed in the next section, covering the axon with an insulating myelin wrapping will also speed up the action potential propagation.

3.3 Active Membrane Properties

The treatment of conduction of axon potentials thus far assumed a non regenerating axon, i.e. the action potentials are propagated, but not "boosted" or "amplified" anywhere along the cable. If this would indeed not be the case, action potentials would never reach their destination, since the response would have died out after already a few mm of traveling. Therefore, Nature uses amplifiers (voltage sensitive channels) that boost the signals along the axon. This implies that if a particular voltage threshold is crossed, a *new* action potential will be generated. Typically, this occurs at a threshold of about $+4-10$ mV. The distance between the "boosters" is partially dictated by the value of the length constant, λ. It is due to this characteristic of the *regenerating axons*, that action potentials can travel long distances.

The actual conduction velocity of action potentials is not only determined by the electrotonic speed. The voltage sensitive channels (in particular the voltage gated sodium channels) have a significant contribution to the action potential propagation velocity, since their opening is not instantaneous once their threshold voltage is reached. It is important to realize that damage over a short nerve segment may already be sufficient to either significantly slow down the conduction velocity or even to block conduction completely. Depending on the particular nerve tract involved, this results in various neurological symptoms, including abnormal sensations or muscle weakness.

3.4 Myelinated Fibers

Axons can be covered with myelin, that is formed by satellite cells. In the central nervous system, these satellite cells are the oligodendrocytes, whereas in the peripheral nervous system the myelin is formed by Schwann's cell. The thickness of the myelin depends on the number of spirals of the Schwann cell membrane. Some nerve fibers are surrounded by only a few turns of the membrane, while others may have as many as 50-100 turns[5]. The opening between the myelin parts is known as the node of Ranvier. This is illustrated in Figure 3.4. The myelin causes a significant increase in the passive (electrotonic) transport velocity. As discussed previously, the conduction velocity is determined by two cable parameters (cf Eq.3.20), the membrane time constant τ_m and the electrotonic distance λ. The shorter the time constant,

[5] Using electron microscopy, it is seen that the myelin is laminated, where each lamella measures about 13-18 nm thick [6].

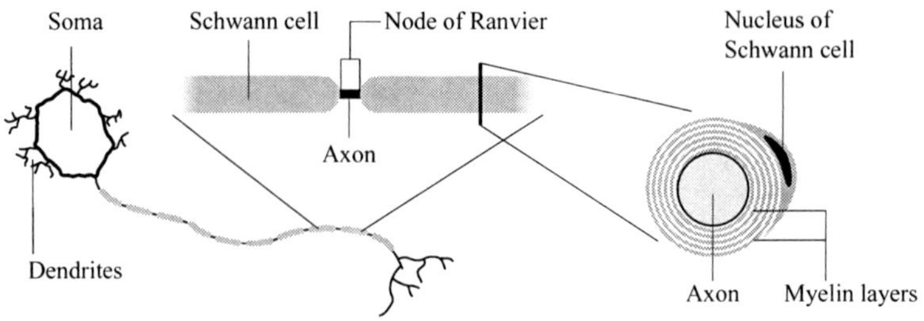

Fig. 3.4 Left: Artist impression of a part of a myelinated fiber. Right: Cross-section of a myelinated fiber. Note the multiple wrappings of the Schwann's cell in this peripheral nerve fiber. This number may vary from a few to one hundred.

the quicker a neighboring region of membrane can be brought to threshold (less charge required) and the sooner this region will fire an action potential. However, the decay of the membrane potential as a function of space should be small, as well, to relatively preserve the shape of the action potential. This preservation of shape ensures that the time needed to cross a particular threshold level relative to the beginning of the change in membrane potential is short. This characteristic is captured by the length constant λ. Therefore r_m must be large (high specific membrane resistance) and the specific axonal resistance r_{ax} should be low, i.e. a/r_{ax} is large (thick fibers, with low specific axonal resistance). To satisfy both a large λ and a small τ_m, implies a large fiber, with small r_{ax}, a large value of the specific membrane resistance, r_m and, to keep τ_m small, a low value of the specific membrane capacitance, c_m. Some rough data on the resistive and capacitative properties of a myelin sheath in comparison with that of a single cell membrane are given in Table 3.1. This high-resistance, low-capacitance combination is indeed realized by the myelinated axons. Since the λ of these myelinated axons is relatively large, it is not necessary to "amplify" the action potentials at a short distance, as is the case for an unmyelinated fiber. Therefore, myelinated fibers allow a relatively large space between those membrane parts where action potentials are generated. These spaces have the length of the myelin sheets, i.e. about 1-5 mm, a value of the order of the length constant for a myelinated nerve fiber. In between, the Ranvier nodes are present, that contain the voltage dependent channels, that are the generator sites of the action potentials.

Table 3.1 Electrical properties of myelin sheath and individual cell membrane. Data from [7].

	specific membrane resistance	**specific membrane capacitance**
	(Ωcm^2)	(F/cm^2)
myelin sheath	10^5	10^{-8}
cell membrane	10^3	10^{-6}

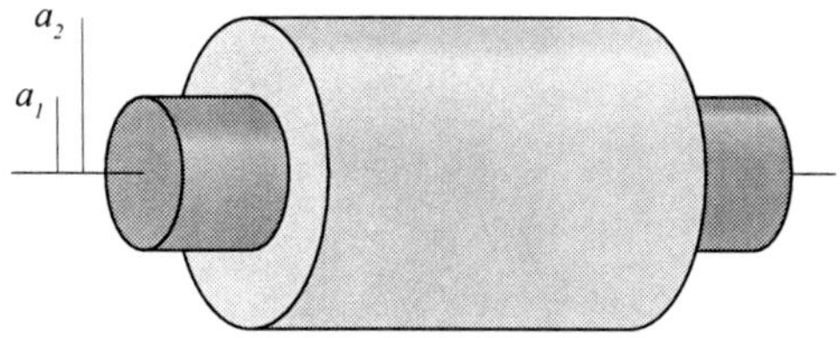

Fig. 3.5 Estimation of the membrane capacitance and the membrane resistance for a myelinated axon with inner diameter $2a_1$ and myelin thickness $a_2 - a_1$. The various membrane layers are not shown.

3.4.1 Estimation of λ and τ for Myelinated Fibers

Conduction along a myelinated nerve is realized by electrotonic conduction along the myelinated segments, and the generation of action potentials at the various nodes of Ranvier. Let us try to find an expression for the space constant λ and the time-constant τ_m. Firstly, we set the inner radius of the axon equal to a_1, covered with myelin, with an outer radius a_2. This is shown in Figure 3.5. To calculate the total capacitance of a myelinated axon, realize that the myelin can be considered as a series of thin (thickness da), concentric cylindrical shells, each with a capacitance

$$c_{\text{layer}} = \frac{\epsilon_0 \epsilon_r 2\pi a L}{da} \tag{3.21}$$

with L the length of the myelin sheet, $\epsilon_0 = 8.85 \cdot 10^{-12}$ F/m the permittivity of free space and ϵ_r the (static) relative permittivity.

Remember, that putting capacitors in series yields for the reciprocal value of the total capacitance C_m the sum (or integral) of the inverses of the individual capacitances. Therefore, we calculate,

$$\frac{1}{C_m} = \int_{a_1}^{a_2} \frac{da}{\epsilon_0 \epsilon_r 2\pi a L}. \tag{3.22}$$

This yields for the total membrane capacitance, C_m

$$C_m = \frac{\epsilon_0 \epsilon_r 2\pi L}{\ln(a_2/a_1)}. \tag{3.23}$$

By using the identity $c_m d_m = \epsilon_o \epsilon_r$, with d_m the thickness of a single layer of myelin, we can also write

$$C_m = \frac{c_m d_m 2\pi L}{\ln(a_2/a_1)}. \tag{3.24}$$

In a similar vein, we can calculate the total membrane resistance of a myelinated fiber. The resistance of a piece of the cell membrane, as function of da is now given by

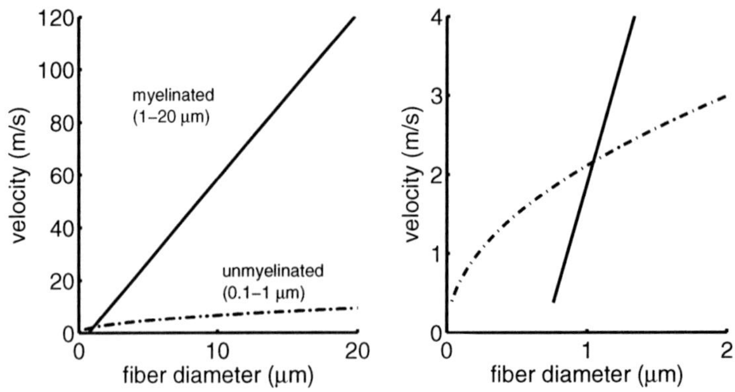

Fig. 3.6 Left: Electrotonic conduction velocity versus diameter for unmyelinated and myelinated fibers. Right: detail for fiber diameters 0-2 μm. In myelinated fibers, velocity scales linearly with the diameter, in unmyelinated fibers velocity squares with the root of the radius. Note, that for very small diameters, unmyelinated fibers conduct faster than myelinated fibers, as shown in the detail at the right. Therefore, myelination speeds up conduction velocity only if the fiber diameter is larger than 1 μm. Myelinated axons less than 1 μm in diameter are indeed rare, and unmyelinated axons larger than 1 μm are similarly rare. The myelinated fiber diameter shown is that including the myelin.

$$dR = \frac{\rho da}{2\pi a L} \tag{3.25}$$

with ρ the resistivity or specific electrical resistance (with unit Ωm). The total membrane resistance R_m is the integral over the thickness of the various cell layers, starting at a_1 and ending at a_2, of the infinitesimally contributions dR

$$R_m = \frac{\rho}{2\pi L} \int_{a_1}^{a_2} \frac{da}{a}. \tag{3.26}$$

This yields for the total membrane resistance R_m

$$R_m = \frac{\rho}{2\pi L} \ln(a_2/a_1). \tag{3.27}$$

Using the identity $\rho = r_m/d_m$, we write

$$R_m = \frac{r_m}{2\pi L d_m} \ln(a_2/a_1). \tag{3.28}$$

For the space constant, λ, we now find, using Eq. 3.13 that

$$\lambda_{\text{myelin}} = a_1 \sqrt{\frac{r_m \ln(a_2/a_1)}{2 r_{ax} d_m}}. \tag{3.29}$$

Since $a_1 \sim 0.6a_2$ in myelinated fibers, λ scales linearly with the diameter. Using again, as an approximation for the velocity, that $v \sim 2\lambda/\tau$, it directly follows that for myelinated fibers the velocity scales linearly with the diameter, as well, since τ can be considered independent of the diameter, as stated above. In sum, therefore, velocity in unmyelinated fibers scales with the root of the radius (or diameter), whereas the velocity in myelinated fibers scales linearly with the radius. This is illustrated in Figure 3.6 for different values of the diameter.

3.4.2 Safety Factor

The typical internodal distance for a myelinated fiber is approximately 100 outside diameters or $200 \cdot a_2$. Note, that this implies that if the fiber becomes thicker, the internodal distance increases, as well. This relationship is also known as the second of Rushton's scaling laws [8], who found this relationship from measurements in cats and kittens. Since velocity scales linearly with the diameter in myelinated fibers, we find for the time difference between the generation of subsequent APs in a myelinated fiber that

$$\Delta t = \frac{\Delta x}{v} \sim \frac{200 \cdot a_2}{c \cdot a_2} = \text{constant}. \tag{3.30}$$

For all myelinated fibers, the internodal transit time is approximately constant, and equals ~ 17 µs. Consider, for instance, a myelinated fiber with outside radius $a_2 = 0.007$ mm. This fiber will have an internodal distance of about $200 \cdot a_2 = 1.4$ mm. Using this internodal transit time of 17 µs, the velocity is approximately 83 m/s. For an axon length of 1 meter, the number of nodes is $1000/1.4 \sim 714$.

The length constant is a relevant dimension for the maximum distance over which electrotonic conduction is allowed before it is necessary to generate a new action potential. The length constant for a myelinated fiber was given by Eq. 3.29. Using typical values[6], we find that $\lambda_{\text{myelin}} \sim 1000 \cdot a_2$. For a fiber with an outside diameter of 4 µm, the length constant is about 2 mm.

As a rule of thumb, the maximum distance an action potential can travel before its amplitude is too low to allow the generation of a new action potential is about 3λ, $3000 \cdot a_2$ or 1500 axon diameters. Given the internodal distance of $200 \cdot a_2$, the safety factor is about 15 nodes. So, remember, if myelin is damaged over a length of more than 15 nodes or 1500 axon diameters, propagation is blocked. Typical axon diameters for human myelinated fibers are in the range 1-20 µm. For an axon with thickness 10 µm, 15 nodes would equal a distance of approximately 15 mm.

[6] Although values differ, a realistic value for the specific membrane resistance (for a single layer of myelin) is $r_m = 2 \cdot 10^3$ Ωcm², and for the specific axial resistance, $r_{ax} = 100$ Ω cm and $d_m = 13 \cdot 10^{-7}$ cm. Further, set $a_1 = 0.6a_2$.

3.4.3 Action Potential Velocity

We have found that it holds for the electrotonic conduction velocity of myelinated fibers that it scales linearly with the radius, a. It also holds, that the internodal distance is a constant times the radius. Assume that the time needed to generate a new action potential is δt. For a myelinated nerve fiber of length L, with n nodes of Ranvier the total time needed for an action potential to cross this distance is given by

$$\frac{L}{v_{\text{electrotonic}}} + n \cdot \delta t = \frac{L \cdot c_1}{a} + \frac{L \cdot c_2}{a}. \tag{3.31}$$

where we used $n \cdot \delta t = c_2 L/a$ and $v_{\text{electrotonic}} = a/c_1$, and assumed that the length of the node of Ranvier $\ll$ internodal distance, and can be neglected. Therefore, we still find a linear relationship between fiber diameter and conduction velocity, since the velocity is the reciprocal value of $(\frac{c_1}{a} + \frac{c_2}{a})$, given by

$$v_{ap} \sim \frac{a}{c_1 + c_2} = \text{constant } \cdot a. \tag{3.32}$$

This is an interesting result. Due to both the linear relationship of the electrotonic conduction velocity with fiber diameter and internodal distance as a linear function of the fiber thickness, the (overall) conduction velocity of myelinated fibers scales linearly with the axon diameter. Note also, that the actual conduction velocity is lower than the electrotonic velocity, due to the time needed to generate new action potentials, that is contained in the second term of Eq. 3.31. We leave it as an exercise for the student to estimate the relative contribution of the ion channel opening times to the action potential conduction velocity.

3.4.4 Saltatory Conduction

In myelinated fibers, the action potential moves fast from node to node by the principles of electrotonic conduction, passing a relatively large distance, made possible by the large value of the length constant, λ. At the internodes, the action potentials are boosted by the voltage-gated channels. This new action potential now "moves" quickly to the neighboring node. This process is known as saltatory conduction (from saltare: to jump) as illustrated in Figure 3.7.

Note, that these action potential amplifiers are present both in myelinated and unmyelinated fibers. In myelinated fibers however, the ion channels are clustered at the internodal parts of the cell membrane. Therefore, for a given length of axon, the total number of channels is much smaller for myelinated than unmyelinated fibers.

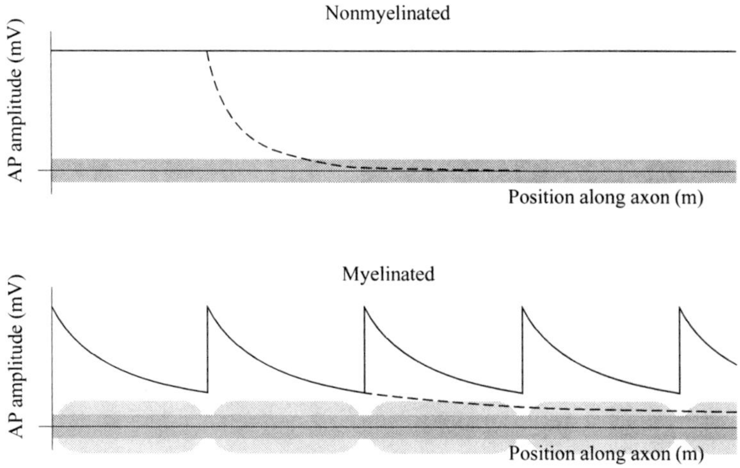

Fig. 3.7 Electrotonic conduction along an unmyelinated fiber (top) and saltatory conduction along a myelinated fiber (bottom). The length constant, λ, is much shorter in the unmyelinated fiber than in the case of a myelinated fiber, shown at the bottom. Amplification needs to be realized over shorter distances in unmyelinated fibers (not shown) than in myelinated fibers.

3.5 Peripheral Neuropathies

A large group of neurological diseases is known as (poly)neuropathies, characterized by a disturbance of the normal functioning of one (mononeuropathy) or more peripheral nerves (polyneuropathy). Three principle groups can be defined, depending on the primary anatomical structure involved: the cell body, the axon or the myelin.

In neuronopathies, pathological changes begin in the cell body, as for instance in amyotrophic lateral sclerosis or sensory neuronopathies[7]. If the axon is primary damaged, an axonal neuropathies is present. If the primary target is the myelin, neuropathies are known as demyelinating neuropathies. A variety of causes may damage the myelin or the axon, including trauma, auto-immune disorders and metabolic disturbances. In the central nervous system demyelination and inflammation along axonal pathways can occur, as well. The most well-known disease characterized by this damage of axonal pathways is multiple sclerosis. This disorder typically starts between ages 20 and 40, and patients may suffer from a variety of neurological disturbances, determined by the location of the affected regions. Particularly common are monocular blindness (due to lesions of the optic nerve: optic neuritis), motor weakness or paralysis (due to lesions of the corticospinal tract), abnormal

[7] Causes of sensory neuronopathies include intoxications (pyridoxine, cis-platinum), infections (Herpes zoster) and some immune disorders.

somatic sensations (due to lesions of somatosensory pathways) or double vision (internuclear ophthalmoplegia, due to a lesion in the medial longitudinal fasciculus). In a later chapter, we will discuss multiple sclerosis in some more detail. Here, we will limit the discussion to peripheral neuropathies. In clinical practice, the adjective peripheral is typically dropped. Therefore, if a physician refers to a neuropathy, it indicates a disease of the peripheral nervous system.

Mononeuropathies refer to abnormal functioning of a single peripheral nerve. Most often, mononeuropathies are caused by compression or entrapment. Peripheral nerves that run superficially or run through relatively narrow structures are therefore most likely to be involved. Compression or entrapment causes local damage of the myelin, resulting in slowing of the conduction velocity or even a complete block. In addition, axonal damage can occur. Most likely, in these situations blood flow to the nerve is compromised, as well. An example is compression of the peroneal nerve where it wraps around the fibular head near the knee. Clinically, these patients suffer from a paresis of the tibialis anterior muscle, resulting in difficulty lifting the foot. If severe, a foot drop may be present. Another example of a mononeuropathy that is quite common is entrapment of the median nerve at the wrist, in the carpal tunnel. The carpal tunnel is a rather small structure in the palm of the hand, formed by various bones of the wrist. The "roof" is formed by a ligament. Nature has designed this passageway rather small, and a branch of the median nerve can easily be compressed in this tunnel, resulting in a carpal tunnel syndrome (CTS). Patients often complain about tingling sensations in their hands and fingers (thumb, index, middle finger and half of the ring finger), often during the night, resulting in wake-ups. Similar symptoms may occur during the day, particularly when performing activities with the wrist bent, such as bicycling or typing. In addition, paresis of the 1st and 2nd lumbricalis and the abductor pollicis muscle may occur. Compression of the median nerve in the carpal tunnel is quite common, with an incidence of about 1% of the population.

Besides mononeuropathies, patients may suffer from a dysfunction of multiple peripheral nerves: polyneuropathies. These will be shortly discussed next.

3.5.1 Axonal Polyneuropathies

Axonal polyneuropathies are characterized by primary damage to the axon. Many causes are known. This includes vitamin deficiencies (B6), diabetes mellitus, medication (chemotherapy for instance), trauma, and intoxications (e.g. alcohol abuse). Besides these acquired neuropathies, there are various genetic axonal neuropathies, as well.

The acquired axonal polyneuropathy most often encountered is the diabetic polyneuropathy. This neuropathy is a very common complication of suffering from diabetes mellitus. At the time of diagnosis, neuropathy is already present in about 7% of the patients. The longer the diabetes exists (in particular type I), the larger the likelihood of developing some form of diabetic neuropathy; estimates range from 10 to 65%. Typically, patients present with symmetrical numbness of the feet, that may be accompanied by spontaneous pain sensations and tingling or abnormal reactions to touch (hyperpathy). During the years, the affected area typically increases, and besides the more distal parts of the lower extremity, the hands may become affected, as well. Although motor symptoms are possible, this is not common. The pathophysiology of the diabetic neuropathy is not understood completely, but the abnormal glucose metabolism is a key factor.

The most well-known hereditary axonal polyneuropathy is Charcot-Marie-Tooth (CMT) disease[8], affecting approximately 1 in 2,500 people. The disease is named for the three physicians who first identified it in 1886 - Jean-Martin Charcot and Pierre Marie in Paris, France, and Howard Henry Tooth in Cambridge, England. CMT is also known as hereditary motor and sensory neuropathy (HMSN) or peroneal muscular atrophy. Patients suffering from HMSN experience dysfunction of both motor and sensory nerves. A typical clinical feature is the weakness of the foot and lower leg muscles. This may result in a foot drop, or, if the weakness is more severe, a high-stepped gait. Foot deformities, such as high arches and hammertoes, can also be present. These are caused by weakness of the small muscles in the feet. As the disease progresses, weakness and muscle atrophy may occur in the hands, as well, resulting in difficulty with fine motor skills. Although sensory nerves are also involved, patients rarely notice significant numbness or pain. The onset of symptoms is most often in adolescence or early adulthood, but presentation may be delayed until mid-adulthood. The severity of symptoms is quite variable in different patients and some people may even never realize they have the disorder. The progression is very gradual. HMSN is not fatal and people with most forms of HMSN have a normal life expectancy.

The structure that becomes (initially) damaged in axonal neuropathies is variable, and in many cases not exactly known. There are many potential targets for injury. This includes the various voltage gated channels, the mitochondria and neurofilaments. Damage to any of these structures will impair normal axonal function. In addition, factors that limit the blood supply to the nerve may also result in axonal damage; this is a mechanism in patients suffering from a vasculitis (inflammation of the wall of blood vessels, often small arteries). Besides the treatment of the underlying cause, if possible, many patients suffering from peripheral neuropathies are treated symptomatically, using drugs. Novel treatments for severe cases are spinal cord stimulation (SCS).

---

[8] Part of the information that relates to CMT was obtained from The National Institute of Neurological Disorders and Stroke, `www.ninds.nih.gov/disorders/`.

3.5.2 Demyelinating Polyneuropathies

Neuropathies with primary damage to the myelin are collectively known as demyelinating (poly)neuropathies. In general, all these diseases slow down the conduction velocity; in extreme cases, conduction is blocked ("conduction block"), with paresis or paralysis as a consequence. A well-known demyelinating neuropathy is the syndrome of Guillain-Barré (GBS).

Guillain-Barré's syndrome is an autoimmune disorder where the myelin sheath of peripheral nerves is damaged in a relatively short time. Patients initially complain about weakness, affecting the lower limbs first, that may rapidly (hours to days) progress in an ascending fashion. Numbness or tingling may be present, as well. As the weakness progresses, the arms may become affected. In addition, the facial muscles may become paretic. The severity of cases varies: some patients experience mild weakness, only, while others may become completely paralyzed and need artificial ventilation. In severe cases of GBS, autonomic dysfunction may occur, as well, manifesting as orthostatic hypotension and cardiac arrhythmias. Most patients recover, but, depending on the severity of the myelin damage, some may endure a protracted recovery and some even remain wheelchair-bound.

3.5.3 Diagnostics in Peripheral Neuropathies

The differentiation of the neuropathies is based on several features, including information from the history (including the presence or absence of diabetes, alcohol intake, speed of symptoms, family history), the neurological examination, laboratory tests, and the findings of the electromyographic examination. This will further be discussed in Chapter 7. For students who wish to learn more about these disorders, a very good website is `http://neuromuscular.wustl.edu/` from the Neuromuscular Disease Center, Washington University, St. Louis, USA.

3.6 Summary

This chapter discussed various issues related to the propagation of action potentials. Electrotonic conduction along axons is presented, including the role of myelin, that allows saltatory conduction. You have a basic understanding of the cable equation, including the biophysical meaning of the length or space constant and the membrane time constant. You understand why conduction velocity in an unmyelinated fiber scales with the root of the axon diameter, and in a myelinated fiber linearly with the axon diameter. In addition, we have presented a few examples of neurological diseases that are characterized by dysfunction at this level: neuropathies.

Problems

3.1. Compare electrotonic conduction and saltatory conduction.

3.2. Define the membrane time constant.

3.3. Explain the concepts membrane time constant and membrane length constant, and their role in the electrotonic conduction.

3.4. What is the role of the myelin sheath in nerve function?

3.5. a. Estimate the length constant of the giant squid axon with a diameter of about 1 mm. b. What is the length constant of a mammalian nerve fiber with a diameter of about 1 µm? Use r_m=2000 Ωcm^2 and r_{ax} = 30 Ωcm for the squid axon and r_{ax} = 125 Ωcm for the mammalian nerve fiber.

3.6. The propagation velocity for an action potential along an unmyelinated, non-regenerating, axon is proportional to the ratio of the length constant to the membrane time constant. Use $\lambda/\tau_m = (a/(2c_m^2 r_{ax} r_m))^{1/2}$ to estimate the velocity in a nerve fiber with thickness of 1 mm. Use r_m=2000 Ωcm^2 and set r_{ax} = 30 Ωcm. Use for the specific membrane capacitance, c_m= 10 nF/mm^2.

3.7. Proof that

$$V_m = V_0 e^{-x/\lambda} + \text{constant} \tag{3.33}$$

is a steady-state solution of Eq. 3.8. Hint: in this case, $\partial v/\partial t = 0$. Assume i_e constant.

3.8. Experimentally, the conduction velocity decreases with decreasing temperature, about 1.2-2.4 m/s per degree C. Can you explain this by changes in the passive (electrotonic) conduction velocity, or do you need an additional contribution from changes in the speed of the generation of the action potentials?

3.9. Describe four clinical features that can be observed in a patient with a severe motor neuropathy. What do you think you will find on evaluation of the stretch reflexes? Do you know an example of a severe motor neuropathy, that can develop in a few days, and is an autoimmune disorder?

References

1. Dayan, P., Abbott, L.F.: Theoretical Neuroscience. Computational and Mathematical Modeling of Neural Systems. MIT Press, Cambridge (2001)
2. Carnevale, N.T., Hines, M.L.: The NEURON Book. Cambridge University Press, Cambridge (2006)
3. Koch, C., Segev, I. (eds.): Methods in Neuronal Modeling From Ions to Networks. MIT Press, Cambridge (1998)

4. Bowser, J.M., Beeman, D.: The book of Genesis. Exploring realistic models with the general neural simulation system (2003), http://www.genesis-sim.org//GENESIS
5. Kandel, E.R., Schwartz, J.H., Jessell, T.M. (eds.): Principles of Neural Science. McGraw-Hill, New York (2000)
6. Snell, R.S.: Clinical Neuroanatomy. Lippincott Williams & Wilkins (2006)
7. Plonsey, R., Barr, R.C.: Bioelectricity: a quantitative approach. Springer, Berlin (2007)
8. Rushton, W.A.H.: A theory of the effects of fiber size in medullated nerve. J. Physiol (London) 115, 101–122 (1951)

Chapter 4
Neural Circuits and Systems

In the previous chapter we discussed the generation of membrane potentials, action potentials and the conduction of action potentials along unmyelinated and myelinated fibers. This chapter will treat the transfer of information and some aspects of information processing at various levels in the central nervous system, in particular between neurons and neuronal networks or circuits. We already shortly touched upon the coding of information in the central nervous system. This information is present at various levels: at the *microscopic* level in individual spike trains, at the *mesoscopic* levels in the electrical behavior of local circuits and at the *macroscopic* level as interactions between neuronal networks. The mesoscopic level is defined as that level where neuronal circuits can be identified with a particular task, for instance language processing, motor control or regulation of particular reflexes. An example of the latter is the vestibulo-ocular reflex, which stabilizes images on the retina when the head is moving. Other examples include systems to perceive pain or touch. The sizes and extent of these various systems differ considerably. There is some compartmentalization, however, and various functions are *localized* to particular regions in the brain. This is exemplified in Figure 4.1. Clearly, the various neuronal circuits need to be connected to transfer information between them. Within the brain, this is realized by the (relatively short) intra- en interhemispheric connections.

These various "levels" of information processing will be discussed in more detail in this chapter. In addition, we will discuss examples of neurological diseases that are caused by "malfunctioning" at these various levels. Some diseases may be primarily caused by a disturbance at a particular level, e.g. the mesoscopic level, but may spread to a larger scale, - an example is epilepsy. We will discuss some of these network characteristics in relation to clinical neurology. This will include paroxysmal disturbances, as may occur in epilepsy. Also, changes in the amount of neurotransmitters available may cause disease; the prototype example is Parkinson's disease, caused by insufficient production of dopamine. First, we will shortly treat some aspects of communication between neurons, the most elementary level of information processing.

M.J.A.M. van Putten: Essentials of Neurophysiology, BIOMED, pp. 53–87.
springerlink.com

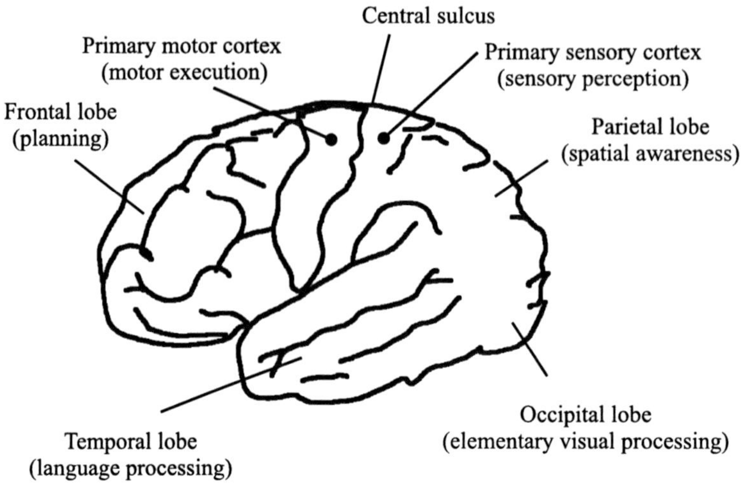

Fig. 4.1 Illustration of various elementary brain systems and an example of a function realized by this particular system (in parentheses). Neurological diseases that are characterized by lesions in the pathways between these systems are sometimes referred to as *disconnection syndromes.*

4.1 The Synapse

The transfer of signals between neurons takes place in a special structure, called the synapse. This is illustrated in Figure 4.2. The synapse shown is a chemical synapse. Action potentials that travel down the axon initially activate ligand gated calcium channels, that allow extracellular calcium to enter the cell. Calcium ions play a vital role in the sequence of events leading to the excretion of a neurotransmitter, as they are needed at the presynaptic site to release the vesicles containing the neurotransmitter. The influx of calcium on its turn eventually causes (through some additional steps) that the vesicles, containing the neurotransmitter, fuse with the axonal membrane, and excrete their contents: the neurotransmitter. The neurotransmitter will cross the synaptic cleft, and interacts with the post-synaptic (ligand gated) receptor, to exert its effect on the receiving neuron. There are many neurotransmitters in the CNS, such as dopamine, glutamate, serotonin and acetylcholine. All these neurotransmitters act on ligand-gated ion channels. As holds for almost any structure or mechanism, be it created by Nature or mankind: it can be malfunctioning at some point. This holds for the various processes that take place in and around the synapse, as well. Disturbances include insufficient release of the neurotransmitter (a pre-synaptic problem), abnormal duration of effects (e.g. caused by an insufficient amount of breakdown enzymes in the synaptic cleft) or abnormal interactions with the ligand gated receptor (a post-synaptic problem). Actual examples include intoxications, that may interfere with the release of neurotransmitters, as occurs in

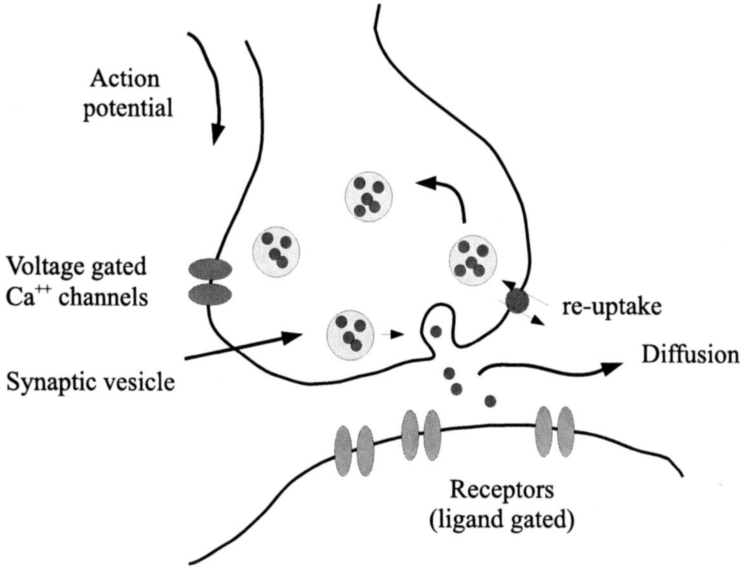

Fig. 4.2 Illustration of a chemical synapse. The effect of the neurotransmitter that is released in the synaptic cleft can be ended by various mechanisms, including diffusion, recycling and degradation by enzymes (esterases), that are present in the synaptic cleft.

Botulism. Some autoimmune diseases destroy ligand gated receptors (as occurs in some forms of myasthenia gravis), thus interfering with the physiological interaction of a neurotransmitter and its receptor. Besides the ligand gated channels, the central nervous system contains many voltage gated channels: these channels are activated by a (sufficient) change in the membrane potential, as extensively discussed in the former two chapters.

4.1.1 Post-synaptic Potentials

The change in potential that occurs at the post-synaptic site due to the release of the neurotransmitter and its interaction with the receptor is known as a post synaptic potential (PSP). The value of the post-synaptic potential is determined by the change in conductance (realized by the interaction of the neurotransmitter with the ligand-gated channel) and the ion-concentration gradient of the ions involved in this particular channel. In general, the membrane voltage will tend to go to the Nernst (also known as the reversal) potential of the ion species that is gated by this channel, as quantified by the Goldman-Hodgkin-Katz equation.

For **excitatory synapses** (for instance a ligand-gated sodium channel), the reversal potential is larger than the resting potential, say +60 mV. Hence, if this synapse becomes active, the membrane potential, V_m, will increase due to the positive current that will flow (positively charged ions) *into* the soma.

For **inhibitory synapses**, however, the situation is slightly different, and three scenario's are possible. This is due to the fact that the reversal potential of inhibitory synapses is typically already close to the resting potential of the neuron. An action potential arriving at an inhibitory synapse will pull the membrane voltage towards its Nernst potential. If the Nernst potential of an inhibitory synapse is more negative than the resting potential, the cell membrane potential will become more negative. If the Nernst potential of the inhibitory synapse is larger (less negative) than the resting potential, the cell membrane potential will *increase*, although the synaptic input is of an inhibitory nature. However, since the mean channel conductance is *increased* by this inhibition, the influence of excitatory input is reduced; in fact, the significant reduced resistance shunts excitatory input the neuron might receive from other synapses. If the reversal potential of the ions involved in the inhibition is near or equal to the existing membrane potential, there will be no significant change in the resting membrane potential after activation of the inhibitory synapse. This is known as "silent inhibition".

We can now differentiate between two classes of input stimuli: stimuli that activate excitatory synapses, and increase the membrane potential (excitatory input, leading to an excitatory post synaptic potential (EPSP)) and stimuli that activate inhibitory synapses. These (typically) reduce the membrane potential (inhibitory input, leading to an inhibitory post synaptic potential (IPSP)), but not in all circumstances, as we have just discussed. Acetylcholine and glutamate are examples of *excitatory* neurotransmitters. The receptors that these neurotransmitters stimulate have a selective conductivity for sodium ions. Glutamate is the most important excitatory transmitter in normal brain function. Nearly all excitatory neurons in the central nervous system are glutamatergic, and is it estimated that over half of all brain synapses release this agent. In clinical neurology, glutamate plays an especially important role, since it may cause neuronal damage if released in too high quantities, causing prolonged excitatory synaptic transmission. This process is known as excitotoxicity. In this process, excessive activation of neuronal glutamate receptors can excite neurons to death. This mechanism is involved in various acute neurological disorders, such as hypoglycemia, trauma and status epilepticus.

An important neurotransmitter that has an *inhibitory* role is gamma aminobutyric acid (GABA); the GABA receptor has a selective conductivity for chloride ions.

The effects of neurotransmitter release can be ended by various mechanisms, including diffusion out of the synaptic cleft, break-down of neurotransmitters by particular enzymes (generally known as *esterases* and re-uptake of neurotransmitters in the terminal axon. The effect of glutamate, for instance, is terminated by uptake into neurons and surrounding glia cells. Although most synapses in the central nervous system are chemical synapses,

Table 4.1 Overview of some characteristics of inhibitory and excitatory (chemical) synapses.

Input	firing likelihood	location	potential	ion involved
Inhibitory	reduced	soma	more negative (*)	Chloride (Cl^-)
Inhibitory	reduced	soma	more negative (*)	Potassium (K^+)
Excitatory	increased	axon hillock	less negative	Sodium (Na^+)

(*) Sometimes less negative if reversal potential is above resting membrane potential. It is also possible, that there is hardly any change in the membrane potential, if the reversal potential of the ion species involved is approximately equal to the resting membrane potential (silent inhibition).

electrical synapses are present in the central nervous system, as well. Electrical synapses allow the transport of signals without the use of a (chemical) neurotransmitter. These synapses, or junctions, are known as "gap junctions", as previously discussed.

4.1.2 Synaptic Strength

There is vast amount of literature that discusses the various mechanisms of changing network properties, including synaptic strength. These processes are relevant in learning and memory formation. Already in 1949, the famous Canadian psychologist Donald Hebb in his 1949 book entitled *The organization of behavior* stated that:

> When an axon of a cell A is near enough to excite cell B or repeatedly or persistently takes part in firing it, some growth or metabolic change takes place in both cells such that A's efficiency, as one of the cells firing B, is increased.

At the time of writing, this statement was rather speculative and Hebb had no techniques available to verify his hypothesis. Nevertheless, his book had a great influence on many researchers. Recent experimental results have indeed shown that changes in synaptic efficiency, do occur, and Hebb's qualitative description is remarkably accurate. The processes in which synaptic strength or efficiency is modulated are commonly denoted as Hebbian learning, in his honor.

Pioneering studies to explore simple memory circuits were performed by Eric Kandel and colleagues. These studies were performed in a large sea snail, *Aplysia*. The circuit they used for their studies was the gill withdrawal reflex. When the snail is quiescent, the gill and the siphon (a fleshy continuation of the gill) is extended from the abdominal region. Water can thus flow over the gills, that serve as the part of the respiratory system. The syphon assists the right positioning of the gill to let aerated water flow across it. In dangerous situations, however, the snail can withdraw the gill and the siphon. This is a defensive reflex, that, in the laboratory, can be initiated by e.g. stroking the syphon with a paint brush. By careful studying this reflex, using various behavioral paradigms, Kandel and colleagues discovered various basic processes related to learning, in particular *habituation*, *sensitization* and *conditioning*.

For instance, they discovered that in habituation, -the process where one learns *not* to respond to irrelevant stimuli -, there is a decrease of the number of synaptic vesicles released for each action potential. Furthermore, if training was longer, fewer sensory neurons had effective connections with motor neurons. The opposite of habituation is sensitization. This is a learning process where there is an *increase* in a response, for instance a reflex magnitude, as a result of stimulation. One of the ions that plays a key role here is calcium. Due to changes in sensory neurons resulting from the effect of modulatory interneurons (that are present in almost all reflex loops), they found that the pre-synaptic calcium channel ultimately remains open for a longer period. As you may recall, pre-synaptic Ca^{2+} is essential for the release of a neurotransmitter. An increased Ca^{2+} concentration in the presynaptic axon will increase the number of released vesicles per action potential. This will, therefore, result in a greater response of the receiving neuron. A very good introductory text that discusses various of these processes in detail is [1].

These studies on *Aplysia* were performed on relatively simple nervous systems. Do similar processes also occur in more complex systems, for instance in the human brain? In fact, they do. However, the mechanisms involved may be quite distinct from those present in *Aplysia*. The overall result, however, a change in synaptic efficiency, is similar. A general term that relates to these changes in the efficiency of the synaptic function is *synaptic plasticity*. Synaptic plasticity can last from only 30 minutes to hours, days or even weeks. The amplifications in the synaptic efficiency are known as long-term potentiation (LTP), reductions as long-term depression (LTD)

The synaptic plasticity depends on the firing rates of pre-and postsynaptic neurons, the synaptic strength itself and time. Hebbian learning is relevant in many physiological and pathological situations. As an example, in patients suffering from long-standing seizures, one of the mechanisms that is most likely responsible for permanent changes in the network architecture, is a change in synaptic strength. This can be mediated by a change in the receptor characteristics and has been experimentally shown to occur in tissue slices of the hippocampus. In this particular experiment, researchers induced "epileptic bursting" in a defined region of the tissue slice, and showed that neighboring parts, that were initially "healthy" changed into abnormal networks, as well [2]. This was caused by a change in the GABA receptor. As you remember, this is a very important receptor that generates inhibitory post-synaptic potentials (with a selective conductivity for chloride ions). Abnormal functioning of this receptor may result in an increased excitation in networks, that may result in seizures. Later, we will discuss seizures and epilepsy in some more detail.

4.2 Firing Neurons

Neurons receive a large number (hundreds to thousands) of inputs from many other neurons. This input architecture consists of a tree like structure called a *dendritic tree.* In addition to the axon-to-dendrite and axon-to-soma junctions, there appears to be a small population of axon-to-axon and dendrite-to-dendrite junctions in the CNS, as well. Each neuron in the central nervous system (brain stem or in the brain) is constantly bombarded by synaptic input from other neurons. Individual EPSPs are only 0.2 to 0.4 mV in amplitude, clearly insufficient to reach threshold. They have to "compete" with the IPSPs; all competing inputs are integrated in the postsynaptic neuron by a process called synaptic integration. In fact, the most fundamental operation in the CNS takes place here: to fire or not to fire. This is *essential decision making.* In this sense, the neuron can be viewed as a local democracy: it counts all the votes at its input (with a particular weight, by the way; this is different from the democracy), and fires if a majority of excitatory inputs is present. A basic question now is: when does the receiving neuron fire an action potential? The receiving neuron will produce an action potential if the weighted sum of these inputs exceeds a particular threshold[1], typically about -55 to -50 mV. In fact, this again is an all-or-nothing law: if and only if the threshold is reached, the neuron will fire; stated differently, the weighted sum of the EPSPs and the IPSPs must be larger than the threshold, V_{th}. Two general mechanisms exist to let the receiving neuron fire. The receiving neuron may fire if the number of simultaneously firing excitatory inputs is sufficiently large to reach threshold - and larger than the sum of the inhibitory inputs. This process is known as *spatial summation.* Spatial summation, therefore, occurs when multiple inputs arrive almost simultaneously from different presynaptic neurons. Since the individual post-synaptic currents[2] will add, it is possible in this way to reach the excitation threshold. An alternative strategy is to increase the firing rate of the excitatory input, without recruiting additional (spatially distinct) input signals: temporal summation. These two processes are illustrated in Figure 4.3.

[1] The classical view of a threshold is a fixed, static value. This is not the case, however. Neurons are dynamical systems (cf. the Hodgkin-Huxley equations that were discussed in the former chapter), and their dynamics is a function of four classes: membrane potential, excitation variables (such as activation of voltage-gated sodium channels), recovery variables (such as inactivation of voltage-gated sodium channels) and adaptation variables (e.g. calcium dependent currents). An excellent treatise on "the threshold" and the dynamics of neurons is Izhikevich [3].

[2] The transmembrane voltage gradient along a length of the cell membrane of the receiving neuron that results from the activation of a single channel with its corresponding post-synaptic currents, is known as a graded potential.

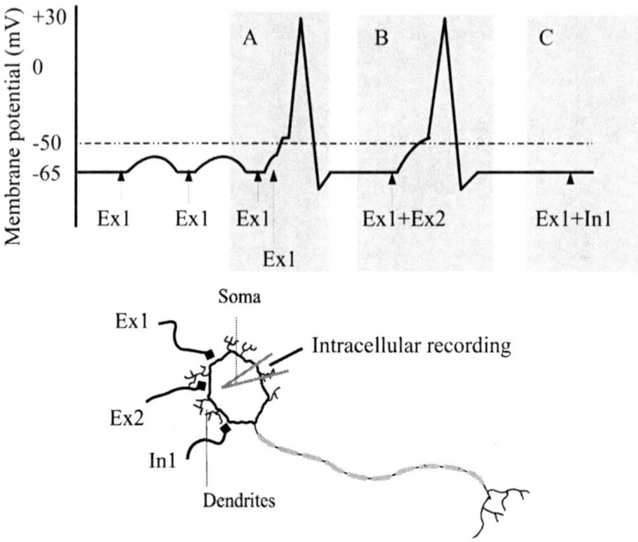

Fig. 4.3 Temporal and spatial summation. In this illustration, the neuron receives three inputs: one inhibitory (In1) and two excitatory inputs (Ex1 and Ex2). A single excitatory input is not sufficient to reach threshold. If the same input neuron (Ex1) fires sufficiently fast, temporal summation realizes that threshold is reached, and an action potential is generated (A). Alternatively, if two excitatory inputs, Ex1 and Ex2, fire at the same moment, spatial summation is sufficient to reach threshold (B). If an excitatory and an inhibitory input are presented at the same time, cancellation may occur (C). The dashed line is the "threshold" for action potential generation.

4.2.1 Integrate-and-Fire Model

An elementary one-compartment neuron model can be used to describe the relationship between the effective synaptic inputs and the firing rate of neurons. The equations for single-compartment models describe how charges flow into and out of a neuron and affect its membrane potential. In this model, we will assume that there are no active membrane conductances (no ion pumps, no Hodgkin and Huxley channels), and there is only a single passive leakage term modeling the entire membrane conductance. At rest, the cell membrane has a membrane voltage of V_{rest}. The leakage term takes care of the finite conductances ($g_L = 1/R_m$) of cell-membranes (recall that there is constant leakage of e.g. sodium ions). Therefore, the neuron model behaves like an electric circuit consisting of a resistor and a capacitor in parallel, as illustrated in Figure 4.4. This very basic integrate-and-fire model only relates to subthreshold membrane potential dynamics. If an external current is applied, it holds, using Kirchhoff's law, that

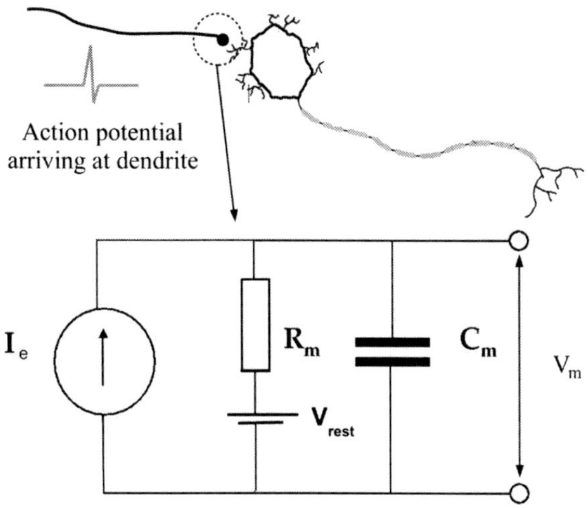

Fig. 4.4 The equivalent circuit for a one-compartment neuron model. The neuron is represented by a single compartment, with constant membrane resistance and capacitance. There is a single channel present, as well. The synaptic input is modeled by the current source, with current I_e.

$$I_e = \frac{V_m - V_{\text{rest}}}{R_m} + C_m \frac{dV_m}{dt}. \tag{4.1}$$

The membrane potential is therefore determined by the linear differential equation

$$\tau_m \frac{dV_m}{dt} = -V_m + V_{\text{rest}} + I_e R_m \tag{4.2}$$

with $\tau_m = R_m C_m$ the membrane time constant. To generate action potentials, we add the rule that if the membrane potential V_m reaches the threshold, V_{th}, an action potential is generated, and the membrane potential is reset to V_{reset}. The membrane potential can now be determined by integrating this equation, and applying the threshold and reset rule for action potential generation.

The time course of the membrane potential of an integrate-and-fire model in response to a constant injected current can now be computed analytically. This yields

$$V_m(t) = I_e R_m \left(1 - \exp(-t/\tau)\right) + V_{\text{rest}} \tag{4.3}$$

where V_{rest} is the value of the membrane potential at time t= 0. It is straightforward to check that this is indeed a solution by substituting it into Eq. 4.2. What we are interested in, is the firing rate of action potentials, given a constant current input, I_e.

From the former equation, we derive the firing rate, f_r, given by

$$f_r = \frac{1}{T} = \left(\tau_m \ln\left(\frac{R_m I_e + V_{\text{rest}} - V_{\text{reset}}}{R_m I_e + V_{\text{rest}} - V_{\text{th}}}\right)\right)^{-1}. \tag{4.4}$$

This expression is valid, only, if $R_m I_e > V_{\text{th}} - V_r$; otherwise, $f_r = 0$. Using the linear approximation of the logarithm $\ln(1+z) \sim z$ for small z, we obtain

$$f_r \sim \left[\frac{V_{\text{rest}} - V_{\text{th}} + R_m I_e}{\tau_m (V_{\text{th}} - V_{\text{reset}})}\right]. \tag{4.5}$$

This equation shows that the firing rate grows linearly with the injected current, I_e for large I_e. So far we have considered an isolated neuron that is stimulated by a single synaptic input, modeled by the external current I_e. In a more realistic situation, the integrate-and-fire model is part of a larger network and the input current I_e is generated by the activity of various presynaptic neurons.

This basic integrate-and-fire model was proposed by Lapicque in 1907, long before the mechanisms that generate action potentials were understood. Although the integrate-and-fire neuron model is very simple, and totally ignores the dynamic processes involved in the generation of the action potential, it can be used in computer simulations to increase our understanding of how neurons interact in large networks [4], like the brain[3]. Also, the integrate-and-fire model illustrates that neurons code their input strength by changing the frequency of the action potential trains they generate. In most cases, however, to simulate large-scale networks, different models are used. This is outside the scope of this book. The interested reader is referred to the literature, for instance [3, 5]. In [5] various Matlab programs are described, as well, to assist in the understanding of various concepts.

4.3 Elementary Circuits

The central nervous system contains various "elementary circuits". Although we previously stated that the mesoscopic level is defined as that level where neuronal circuits can be identified with a particular task, this still leaves room for various interpretations. Here, we will present various elementary circuits that are relevant for clinical neurology.

4.3.1 Motor Control

Movement and motor control is a very important task of the nervous system. In fact, almost all our output is realized by particular motor functions, either

[3] Although some concern the integrate-and-fire model a spiking model, this is not the case. There is no actual mechanism in the IF-model that generates a spike.

directly: locomotion, speech, writing, playing music, or indirectly: putting make-up on, writing a letter or an e-mail, etc. We will not discuss in detail the anatomy of the various motor pathways; this is treated in your neuroanatomy course. Therefore, we will assume you are sufficiently familiar with various concepts. Some aspects will be summarized here for convenience, however.

The motor cortex (cf Figure 4.1) is where the pyramidal tract starts. The motor cortex can be divided into two major parts: the primary motor cortex and the secondary motor cortex. The primary motor cortex (M1) connects to the lower motor neurons, located in the anterior horn of the spinal cord (and brain stem). The lower motor neurons, on their turn, connect to the voluntary muscles. The secondary motor cortex contains the posterior parietal cortex, the premotor cortex and the supplementary motor cortex (SMA). The SMA is important for the planning of movements and the coordination, while the premotor cortex regulates motor guidance. The posterior parietal cortex has a role in assessing the context in which movements are made. The parietal cortex processes somatosensory, proprioceptive and visual inputs, necessary for e.g. defining the position of the body and the targets in space. These processes take place before the actual movements are made with involvement of the premotor and the motor cortex.

The upper motor neurons are cortical pyramidal cells, whereas the motor neurons in the spinal cord are alpha motor neurons. These alpha motor neurons supply the skeletal muscles, and are relatively very large (soma 50-100 μm). The axons of the pyramidal cells that originate in the primary motor cortex and the alpha motor neurons can be quite long, as well, up to approximately 1 meter.

In clinical neurology, it is very important to be able to differentiate between upper motor neuron diseases and a lower motor neuron disease, since these two conditions both present with quite different symptoms and signs and have (in most cases) a very different cause and prognosis. An upper motor neuron disease is defined as a disease that affects the motor pathway that starts in the primary motor cortex, and ends at the lower motor neuron, located in the anterior horn.Therefore, any interruption in the corticospinal tract is clinically defined as an upper motor neuron disease. Examples include strokes (affecting the motor cortex) and spinal cord injury. In an upper motor neuron lesion, the main complaint is often not loss of muscle strength, but rather motor control. Patients may, for example, complain about clumsiness, as manifested by the inability to perform fine motor tasks (such as unbuttoning a shirt). Muscle tone is increased, as opposed to a lower motor neuron disease. Stretch reflexes are typically *increased*, as well. If the lower motor neuron is damaged, a patient suffers from a flaccid paresis or paralysis, loss of muscle tone and a reduction or absence of stretch reflexes. Fasciculations (visible twitching of small groups of muscle fibers) and wasting may occur, as well. Table 4.2 presents an overview.

Table 4.2 Overview of some key differences between an upper and a lower motor neuron (MN) disorder.

Feature	Upper MN	Lower MN
Reflexes	increased	diminished or absent
Muscle tone	increased	decreased
Wasting	moderate	severe
Abnormal reflexes	possible	no

The Stretch Reflex

A reflex is a response to a perturbing stimulus that acts to return the body to homeostasis: feedback. The neural pathway that a nerve impulse follows to realize a particular form of feedback mediated by the nervous system is a *reflex arc*. The components of a reflex arc are: i) a sensor, or receptor, that is part of a sensory neuron, ii) an integration center (one or more synaptic structures in the central nervous system) and iii) an effector. Often, the effector is a motor neuron, that conducts a nerve impulse along an efferent pathway to a muscle.

The prototype example of a reflex is the stretch reflex. The stretch reflex is a muscle *contraction* in response to stretch within the muscle. This is a spinal reflex, that protects muscles against those increases in length which may damage muscle fibers. As you will remember from the neuroanatomy course, this is a monosynaptic reflex, that is modulated by upper motor neuron input. In the stretch reflex, the muscle spindles are the sensors, that transmit stretch of the muscle towards the lower motor neuron, that on its turn connects again to the muscle in which the spindle is located. Examples include the patellar (knee-jerk) reflex, the achilles stretch reflex, the biceps reflex and the triceps reflex.

In clinical examination, these stretch reflexes are very important to differentiate between upper and a lower motor neuron diseases and to detect asymmetries. A normal stretch reflex indicates integrity of the anatomical connections between stretch receptors, the lower motor neuron in the anterior horn and the connections from this neuron back to the muscle. Modulation of the reflex arc is by input from higher centers in the brain, mediated through the corticospinal, reticulospinal, tectospinal, rubrospinal and vestibulospinal tracts. This is sketched in Figure 4.5.

Renshaw neurons, that are located in the spinal cord, receive their excitatory (cholinergic) input from collaterals of the alpha motor neurons, which in their turn, synapse on the lower motor neurons through inhibitory (glycinergic) synapses. These Renshaw cells provide feedback on the lower motor

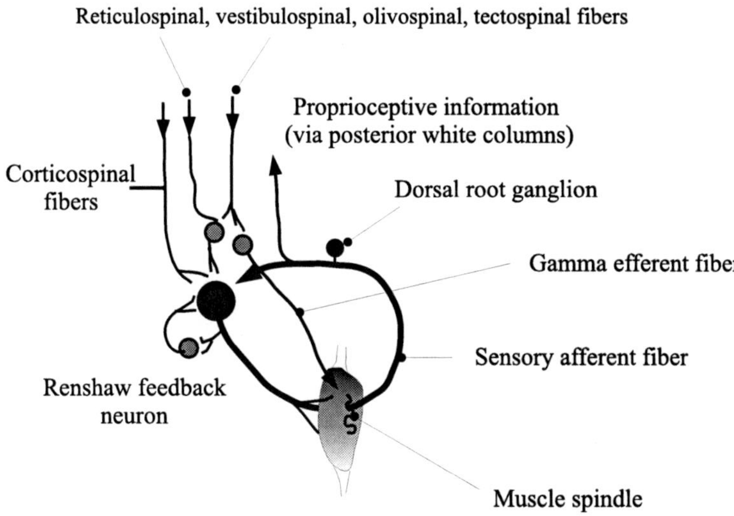

Fig. 4.5 The monosynaptic reflex arc (bold line) and modulating input from higher brain centers by the corticospinal, reticulospinal, vestibulospinal, olivospinal and tectospinal fibers. The Renshaw feedback neurons are indicated as well. Gamma efferent fibers regulate muscle tone. Redrawn and modified from Figure 4.23 in [6].

Table 4.3 Some clinically relevant stretch reflexes and the roots involved.

Muscle (spindle)	reflex arc via spinal root
Biceps	C5/C6
Triceps	C7
Quadriceps	L4
Gastrocnemius	S1

neuron, inhibiting their activity. This is a classic example of negative feedback, or *recurrent inhibition*[4].

If a stretch reflex is reduced, the lesion is typically located in the reflex arc itself. For instance, if a peripheral nerve is damaged (e.g by direct trauma, compression or an infection), the stretch reflex is low or absent. Due to the modulation of the reflex arc by upper motor neuron input (typically reducing the feedback gain of the reflex arc in physiological conditions), upper motor neuron lesions will result in exaggerated stretch reflexes: hyperreflexia. An overview of important stretch reflexes and the roots involved is shown in Table 4.3. Note, that besides stretch reflexes, several other reflexes are known,

[4] Victims of strychnine poisoning will suffer from severe muscle contractions, since strychnine is a glycine receptor blocker; this results in a reduction of the tonic inhibition of the alpha motorneurons by the Renshaw cells.

including e.g. cutaneous reflexes. The most familiar cutaneous reflex in clinical neurology is the plantar reflex: by striking the sole of the foot, the toes will curl, which is a physiological phenomenon. This reflex is also modulated by upper brain centers. Damage to this system (for instance caused by a hemispheric stroke) may result in an upward movement of the toes. This is a Babinski sign, an example of a pathological reflex. In addition to the clinical examination, various techniques exist to evaluate reflex loops, as well, e.g. by using electrical stimulation of the nerve. For further reading, see e.g. [6, 7].

Basal Ganglia

The basal ganglia are a collection of neural circuits, located deep to the white matter of cerebral cortex, that have a major role in movement control. In addition, the basal ganglia are involved in cognitive processing and behavioral regulation. The basal ganglia themselves are a conglomeration of distinguishable interconnected networks. They are called the:

- striatum, consisting of caudate nucleus and putamen;
- globus pallidus, internal, GPi, and external segment, GPe, and the ventral pallidum, VP;
- subthalamic nucleus, STN;
- substantia nigra (pars compacta, SNc, and pars reticularis, SNr).

The connections between the various circuits that constitute the basal ganglia are illustrated in Figure 4.6.

The basal ganglia play a key role in controlling the output of the thalamus to the cortical areas that are involved with movement. There are different connections towards and from the basal ganglia, that we will shortly discuss. Much more details can be found in standard neuroanataomy textbooks.

The striatum and the subthalamic nucleus receive input from various cortical areas (pre-frontal cortex, parietal and temporal cortex) and the substantia nigra (pars compacta). The outputs of the basal ganglia originate at the GPi, SNr and the VP and project mainly to the thalamus (and via the thalamus towards various cortical areas) and the formatio reticularis. In this way, a feedback loop is created from the cortex and substantia nigra towards the basal ganglia and from the basal ganglia towards the thalamus back to various cortical structures. The striatum is the main recipient of input to the basal ganglia, and is connected to the thalamus (via the GPi/SNr, see Figure 4.6) by two different pathways. The direct pathway connects from the striatum directly to the GPi and SNr, and inhibits these two circuits. The alternative route towards the GPi is indirect: first, there is a connection with the striatum and the external segment of the globus pallidus (GPe). From here, there is connection with the subthalamic nucleus (STN), that on its turn connects to the GPi and SNr, and activates these circuits.

The direct pathway is modulated by dopamine 1 (D1), the indirect pathway by dopamine 2 (D2) receptors. The dopamine activates the direct pathway,

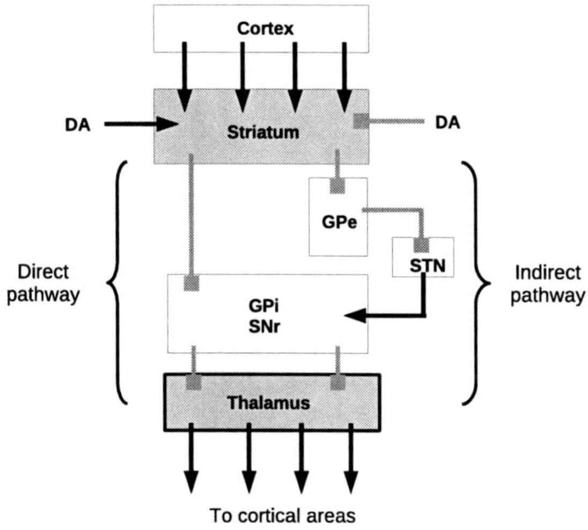

Fig. 4.6 Schematic of the connections from the cortex towards the basal ganglia. Light gray arrows with boxed endings indicate suppression of the target, black arrows indicate stimulation of target structure STN=subthalamic nucleus. SNr=substantia nigra, pars reticularis, GPe=globus pallidus externa. GPi=globus pallidus interna.

but inhibits the indirect pathway. The GPi/SNr, on its turn, sends inhibitory output to the thalamus (ventral lateral nucleus). Selective activation of the cortical areas that receive input from the thalamus is mainly controlled by the direct pathway; undesired collateralmotor activity is inhibited by the indirect pathway. This balance between these two pathways is essential for the normal functioning of the basal ganglia and their corresponding output. Changes in this circuit result in various disturbances, including motor control. Parkinsons disease is the prime example of a disease that is caused by an abnormal ratio in the activity of these two signaling pathways, that will be discussed later in this chapter.

4.3.2 Language

In order to communicate language by sound waves, three essential systems need to function properly. Firstly, phonation is needed, i.e. the generation of an air flow passing through the vocal cords that is subsequently modulated by actions of the tongue, lips and facial muscles. Disorders in phonation are collectively referred to as dysphonias. Secondly, articulation is necessary: the sound wave is modulated by various speech muscles, in particular the tongue. *Dysarthria* is a speech disorder that is due to a weakness or incoordination of these muscles. Causes include stroke, degenerative diseases, myasthenia and ALS (Lou Gehrig's disease). And finally, language needs to be generated in the brain. Acquired disorders of language processing are known as dysphasias, sometimes referred to as aphasias. Review the various structures needed for proper language production, e.g. from your neuroanatomy course.

4.3.3 Memory

There are several different types of memory, each of which is predominantly located in a different part of the brain, including the hippocampus, amygdala and basal ganglia. Declarative, or explicit, memory relates to facts, dates and events. The hippocampus is an essential and critical structure involved. Procedural, or non-declarative/implicit, memory relates to how to perform an act. Examples include playing the piano or riding a bicycle. Here, basal ganglia (dorsal striatum, caudate-putamen) are involved.

In addition to the declarative and procedural memory, there is memory related to a form of learning based on the tendency of certain natural events (e.g. food presentation) to evoke an involuntary responses (e.g. salivation): Pavlovian conditioning. In this process, a conditioning stimulus (CS) is presented at moments where the organism is presented an unconditioned stimulus (US) that is typically related to a particular unconditioned response (UR). As a CS, a tone can be used, while presenting food (US) that is related with salivation (UR). Another example, that you can easily verify at home, is to make a sound, while delivering a pulse of air to the eye (US), causing an eyeblink as the UR. If this procedure is repeated a number of times, the organism is likely to show a conditioned response (CR) if only the CS is presented. For instance, salivation or an eyeblink as the CR in response to a sound (the CS). Learning an association between a CS and a CR is called Pavlovian or classical conditioning, or sometimes "associative learning".

Pavlovian conditioning is an automatic process, and control or voluntary participation is not necessary. In fact, you do not even need to be aware of the association in order to learn by Pavlovian conditioning. The amygdala, located deep in the temporal lobe of the brain, is a relevant structure involved in Pavlovian conditioning. Therefore, if you experience a particular state of mind while listening to a piece of music, this association is most likely mediated by the amygdala.

If a memory event is stored in the brain, it is initially in a labile state, and easily "removed" by external events. The classical example of such an external event is a mechanical stress factor, that can be present in patients suffering from head trauma. If this is sufficiently severe, typically with loss of consciousness, part of the just stored memories are erased. This phenomenon is known as a retrograde amnesia. This results in the loss of recent memories, only, ranging from minutes to hours, depending on the severity of the insult. Often, the process of acquiring new memories is temporarily disturbed, as well (anterograde amnesia), and in these circumstances it is very typical for patients to ask the same questions over and over, e.g. "Please could you tell me what happened?", even if properly answered.

Chronic disturbances in memory function are collectively known as amnesic syndromes. In clinical neurology, the prototype example is Alzheimer's disease. This will be discussed in the next section, amongst other diseases.

4.4 Clinical Examples

The final section of this chapter presents various neurological diseases characterized by disturbances at the circuit level or the connection between various circuits. Disease entities characterized by dysfunction of these connections were sometimes referred to as disconnection syndromes [8].

4.4.1 Parkinson's Disease

The most well-known movement disorder is Parkinson's disease. Parkinson's disease is the second most common degenerative disease of the nervous system, Alzheimer's disease being the leader. This disease was named after the English Medical Doctor, James Parkinson, who in 1817 was the first person to describe these symptoms as "the shaking palsy". The clinical symptoms in patients suffering from Parkinson's disease are referred as Parkinsonism. The essentials of Parkinsonism are (the acronym is TRAP):

- Tremor at rest: involuntary trembling of the limbs (about 4-7 Hz), where the first and second finger of the hand may show a characteristic pill-rolling tremor[5];
- Rigidity of the extremities and neck;
- Bradykinesia or Akinesia: Lack of automatic movement (e.g. arm swing) or slowness in initiating and maintaining movement; often with minimal facial expressions ("masking" or "poker face");
- Postural instability: Characteristic bending or flexion of the body, associated with difficulty in balance and disturbances in gait.

In a more advanced clinical stage, walking is characterized by short steps, a stooped posture, and a paucity of associated movements, such as arm swing. Typically, patients with Parkinson's disease present initially with asymmetric symptoms, e.g. a tremor of the left arm, only. The increased muscle tone on clinical examination is further characterized by a cogwheel rigidity: on passive bending and flexing of an extremity, there are typical fluctuations in the muscle tone, often increased by *active* movements of the contralateral extremity. This is easily verified on neurological examination by asking the patient to make a circular movement with the contralateral arm, while the muscle tone of the other arm is evaluated. Patients may also complain about difficulties on turning in bed, as another manifestation of the increased muscle tone and bradykinesia.

[5] In some countries, e.g. the Netherlands, it is also referred to as a "money-counting" tremor.

Fig. 4.7 Illustration of the hallucinations that a patient suffering from Parkinson's disease may perceive. This patient was an amateur-artist, that suffered from visual hallucinations for a couple of weeks after a minor change in the medication. The water color painting shows "Droodles", as she called it: various distorted shapes from animals and humans that partially merge. She made a few dozen of these paintings, that she showed to me during a consultation. In fact, she told, she liked this period of hallucinations. They were non-frightening and were rather stimulating for her artistic talents. Upon my request, she was so kind to provide me this picture. The original painting is in color, it was transformed to a BW-image using GIMP. Reproduced with permission from the patient-artist.

In addition, many other symptoms can be present, such as hypophonia, drooling (caused by a weak and infrequent swallow), micrographia, and many non-motor symptoms, including autonomic dysfunction, depression and dementia. Sometimes, hallucinations occur as a side effect of dopaminergic medication. An example of these hallucinations is shown in Figure 4.7.

The defects in motor function are due to the progressive loss of dopamine from the dopaminergic neurons in the substantia nigra (pars compacta), that project to the striatum. At birth, the substantia nigra contains about 400-thousand dopaminergic cells, falling to 250-thousand by the age of 60 [9]. There is a parallel loss of dopamine in the striatum [10]. When the cell population or the dopamine content reaches 20% of the youthful levels, signs of Parkinsonism appear [11]. In Parkinson's disease (PD), decreased

Table 4.4 Overview of Parkinson-plus syndromes. All these conditions share Parkinsonism, but the additional features differ, as indicated.

Parkinson-plus syndrome	additional features
Multiple system atrophy (MSA)	cerebellar ataxia, autonomic failure, pyramidal tract disorder
Progressive supranuclear palsy (PSP)	failure of eye movements (especially downward gaze) pseudo-bulbar palsy, dementia, early postural instability
Corticobasal degeneration (CBD)	cognitive impairment, myoclonus, alien limb
Dementia with Lewy bodies (DLB)	early dementia, hallucinations

striatal dopamine causes increased inhibitory output from the GPi/SNr. This increased inhibition of the thalamocortical pathway suppresses movement, resulting in hypokinesia and bradykinesia. The changes in the excitatory and inhibitory levels of the direct and indirect pathways in a Parkinsonian patient is illustrated in Figure 4.8. Compare this with the previous illustration, Figure 4.6, that shows the physiological situation. Note, that all these changes eventually result in a diminished output from the thalamus to the cortical areas. Besides Parkinson's disease, various other diseases exist that may cause

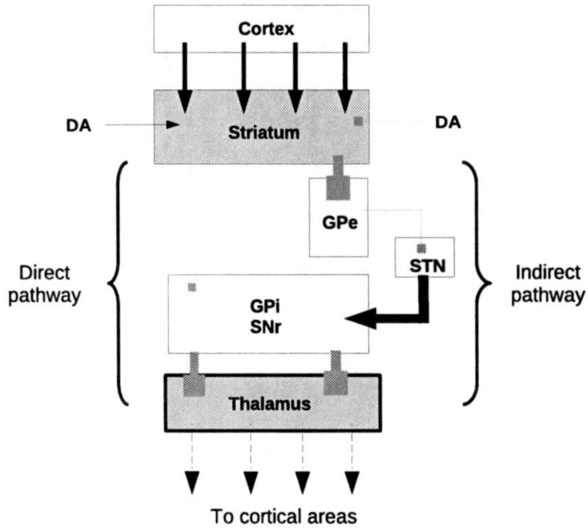

Fig. 4.8 Dopaminergic pathways of the human brain in Parkinson's disease. Light gray arrows with squared endings indicate suppression of the target, black arrows indicate stimulation of target structure. SN=substantia nigra. STN=subthalamic nucleus. GPe=globus pallidus externa. GPi=globus pallidus interna.

Parkinsonism. In most cases, these diseases not only feature the classical symptoms of Parkinson's disease, but additional features, as well. Therefore, they are often referred to as Parkinson-plus syndromes. Examples of these diseases, with their accompanying features, are presented in Table 4.4.

Also, particular drugs may cause Parkinsonism, such as neuroleptics, e.g. Haldol.

Unfortunately, at present, the cause for the progressive deterioration of the neurons in the substantia nigra is not known. Hence, Parkinson's disease is often referred to as idiopathic Parkinson's disease (IPD). For a much more in depth treatise of Parkinson's disease and related disorders, see e.g. [12].

Diagnosis

The diagnosis idiopathic PD is based on the medical history and the neurological examination. CT and or MRI scans are often made to rule out non-idiopathic PD. At present, there are no blood or laboratory tests that can prove with certainty that a patient is suffering from PD. Various tests, however, increase the likelihood of PD. For instance, 18F-6-fluorodopa positron emission tomography (18F-dopa PET) can be used as a marker of presynaptic dopamine storage capacity. Treatment of Parkinson's disease includes the use of dopaminergic medication, dopamine agonists, and deep brain stimulation (DBS).

Epidemiology

Parkinson's disease is encountered all over the world, with a prevalence between 1 to 2.5 per thousand. Although reported at all ages, PD is quite rare in people younger than 40. A famous young Parkinson patient is Michael J. Fox, in whom the first symptoms appeared when he was only 36 years old[6]. In most cases, however, symptoms begin around the age of 60, with a slowly progressing course.

4.4.2 Epilepsy

Epilepsy is the prototype example of a paroxysmal neurological disease. Paroxysmal diseases are conditions characterized by an intermittent occurrence of the symptoms or signs. This includes diseases as migraine or psychiatric diseases - such as a manic-depressive illnesses-. Here, we will focus on epilepsy, however. Patients with epilepsy experience *seizures*. These seizures occur seemingly randomly, and in many patients long periods (weeks, months or longer) may pass in which clinical signs are not present.

A seizure is a transient occurrence of signs and/or symptoms due to abnormal synchronization of neuronal activity in the brain. An additional

[6] See "The Michael J. Fox Foundation for Parkinson's research" for more information. `http://www.michaeljfox.org/`

Table 4.5 Overview of clinical differentiation of seizure types.

Seizure Type	consciousness	Cortex involved
Partial	preserved	one hemisphere, small part
Complex partial	reduced	one hemisphere, larger part
Secondary generalization	absent	both hemispheres
Primary generalized	absent	both hemispheres

aspect is, that the cerebral cortex is involved. This differentiates seizures from other "synchronization disorders", that find their origin in abnormal oscillatory behavior elsewhere in the brain, such as tremor in Parkinson's disease (basal ganglia). The signs and/or symptoms of the pathological synchronization (typically, too strong, but this is not invariably so) can be reflected in various different ways, including changes in mental behavior, experience, or motor behavior. A good clinical working definition of epilepsy is that a patient suffers from epilepsy if his/her brain has an increased tendency to generate seizures, and at least one seizure actually occurred.

There are various ways seizures can be categorized, one of which relates to the spatial distribution of the abnormal synchronization, i.e. partial seizures, partial seizures with secondary generalization and primary generalized seizures. In partial seizures, the region of abnormal synchrony is confined to a relatively small part of the brain, for instance the motor cortex or temporal lobe, only. In these patients, consciousness is essentially preserved, and the clinical manifestations are typically limited to a single disturbance, for instance involuntary movements of the arm, or a transient dysphasia[7]. If a seizure initially starts focally, the neurons involved may recruit neighboring neurons, and the amount of neuronal tissue involved in the pathological synchrony may increase. If by this process consciousness becomes reduced, these seizures are known as complex partial seizures. Impaired consciousness implies decreased responsiveness and awareness of self and surroundings. In these situations, automatisms may occur, as well. Automatisms are quasi-purposeful motor or verbal behaviors, such as lip smacking, chewing or swallowing.

In some circumstances, the activity spreads even more, including the contralateral hemisphere. These seizures are classified as partial seizures with secondary generalization. In these situations, consciousness is typically absent. Clinical manifestations are now often characterized by bilateral tonic-clonic movements. An overview of this differentiation in seizure types is presented in Table 4.5. Although anyone may suffer from a single seizure in

[7] Here, history taking is essential, since transient neurological deficits are also encountered in patients suffering from transient ischaemic attacks (TIA). Important clinical information to differentiate between a seizure disorder and a TIA is the speed of occurrence of the symptoms. In patients with TIA's, this is typically within a few seconds, while in seizure disorders symptoms may appear more gradually, especially in focal seizures.

his lifetime, patients with epilepsy shown an increased tendency to generate seizures. About 1% of the world population suffers from epilepsy, and the causes are numerous, including channelopathies, as discussed in the former chapter, trauma, encephalitis, metabolic disorders, and stroke. As a rule of thumb: any structural damage to the brain increases the likelihood of developing seizures[8]. These different causes allow a second way to classify seizures, taking the (presumed) etiology into account. If the etiology is known, the seizures are classified as symptomatic seizures. A stroke patient, therefore, may suffer from symptomatic, partial seizures. If the etiology is not known, seizures are classified as *idiopathic.* A third division is based on the part of the cortex that is involved, for instance frontal lobe seizures or temporal lobe seizures.

There are various challenges related to epilepsy. We will first discuss some challenges from the clinical point of view, in particular related to the diagnostic process and the treatment of patients with epilepsy. Hereafter, we will shortly touch upon the issue of epileptogenicity: which process are involved in making the brain epileptogenic?

Clinical Aspects of Epilepsy

If a patient suffered from a single seizure, what is the likelihood he/she will have epilepsy? Note, that a single seizure does not imply that a patient suffers from epilepsy. As stated previously, epilepsy is a neurological condition where the likelihood of getting seizures is significantly increased. Although a single seizure may occur in about 5% of the population, in patients with epilepsy there is a substantial risk of *repeated seizures.* In these patients, therefore, the neurologist is faced with the question: is this an incidental seizure or is this a manifestation of epilepsy?

The work-up typically includes a thorough history taking, neurological examination and often computed tomography or MRI-scanning of the brain. An account of an eye-witness may be highly relevant, as well, since this may provide relevant information about how the seizure evolved in time. For instance, what was the first thing noted about the patient? Did the seizure start with difficulty in talking or involuntary rhythmic movements? All this will provide clues about the region of the brain where the seizure originally started, which is relevant for classification and potential treatment.

Part of the diagnostic work-up is focused to investigate if a structural abnormality can be held responsible for the seizure. If this is found to be the case, the seizure has most likely been symptomatic, which makes the diagnosis of symptomatic epilepsy almost certain.

[8] The processes involved here are only partially understood. In post-stroke epilepsy, for instance, -occurring in about 7% of stroke patients- glial damage may be relevant. Glial tissue plays in important role in the buffering for the extracellular potassium. Reducing the potassium buffering capacity of the glial cells results in an increase in extracellular potassium, and a resting membrane potential closer to the firing threshold.

The electroencephalogram (EEG) is a very important technique to evaluate the chance of suffering from epilepsy, as well. In fact, the EEG in patients with epilepsy may show particular, abnormal, wave forms, such as spikes or spike-wave discharges. These abnormal wave forms are "signatures" of the brain's tendency to generate seizures, and indicate an increased likelihood of pathological synchronization between neurons or neural networks. Details of the role of the EEG in epilepsy will be treated in a separate chapter. Another challenge is how to treat the patient, if the diagnosis epilepsy is sufficiently certain. The goal of the treatment is to reduce the likelihood of seizures, preferably to a level equal to the general population. Basically, there are three possibilities: treatment with medication (anti-epileptic drugs), neurostimulation, and treatment by resection of the epileptic focus. It depends on various factors which choice is made, including the number and severity of seizures the patient will experience without treatment, the underlying cause and the age. Neurostimulation is discussed in a later chapter.

As you recall, sodium channels are essential for the neuron to generate action potentials. Various anti-epileptic drugs act on these sodium channels, functioning as a blocker. An example is diphantoine. Other drugs may act on inhibitory receptors, for instance the GABA-receptor. The most familiar drugs that act here are the benzodiazepines[9]. In about 70% of the patients, treatment with medication is satisfactory. In the remaining 30% of the patients, seizures are not well controlled. In this group, in the case of focal seizures, with a well-defined cortical generator of the seizures, surgery can sometimes be performed. This is often the case in some forms of temporal lobe epilepsy. Sometimes, patients present with a status epilepticus. This is a condition characterized by persistent seizure activity, where (in the case of generalized seizures) consciousness is absent for more than 30 minutes. If this condition cannot well be controlled with standard medication, patients with a status epilepticus are often treated in the Intensive Care Unit, including medically induced coma. A very good book that discusses many aspects of status epilepticus, including pathophysiology and treatment, is [13].

Hebbian learning may be relevant in those patients suffering from long-lasting or recurrent seizures, and maybe even more so in children. It is generally acknowledged that neural plasticity (the ability of the brain to "shape" circuits) is a function of age. In general, the younger the individual, the more plastic the brain is. In Hebbian learning, it is postulated that the more a particular circuit is exposed to a fixed input stimulus, the easier it becomes for the circuit to show the associated response. This implies, that, as time progresses, the intensity or even the specificity of the input stimulus may change, while the response remains essentially unchanged. In some circumstances, patients may suffer from long-standing seizures or the repeated occurrence of epileptiform discharges. It is likely, that these events induce various changes in network properties (by Hebbian learning) that make the occurrence of

[9] Benzodiazepines are not only used for the acute treatment of seizures, but also for sedation or (temporary) treatment of some sleep disorders.

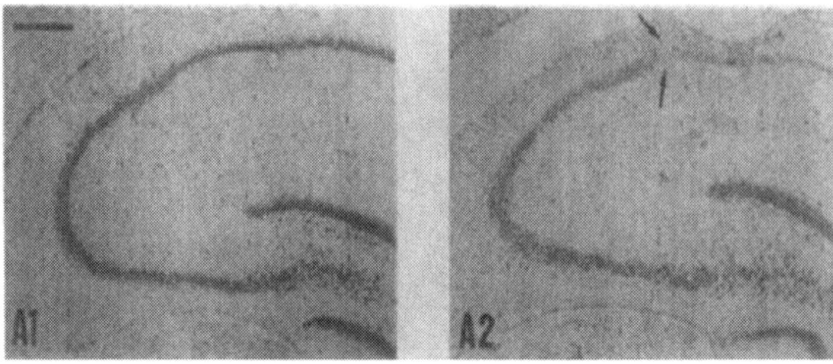

Fig. 4.9 Patterns of cell loss in different areas of the hippocampal formation. A1: The normal aspect of the CAI, CA3, DG, and Hilus in a control rat. A2: There is extensive destruction in CA1 (arrows) in an epileptic animal (36 days after status epilepticus). Illustration from E.A.M. Luiz et al. 'Circuit Mechanisms of Seizures in the Pilocarpine Model of Chronic Epilepsy: Cell Loss and Mossy Fiber Sprouting' Epilepsia, 34(6); 985-995, 1993. Blackwell Publishing, Oxford, England ©1993.

future seizures more likely. This suggests that prompt and rigorous treatment of seizures, and even epileptiform discharges without clinical manifestations, especially in children, is beneficial for the long-term prognosis of patients suffering from epileptic seizures [14]. Additional mechanisms are most likely relevant, as well, for instance the induction of cell death after long-standing seizure activity.

Basic Issues in Epilepsy

Many basic issues in epilepsy are unresolved. In fact, in most patients with epilepsy, it is not fully understood why their neurons show the increased tendency to synchronize pathologically. Related issues include the termination of seizures: how seizures end is often not clear. The mechanisms involved in epilepsy can be studied at various anatomical scales. At the level of the individual neuron, ion channels and synaptic proteins may be involved. For instance, a rather benign form of epilepsy is characterized by a dysfunctional potassium channel, with clinical manifestations at a very young age (months). If these children grow older, this channel is replaced by the adult form of the channel, that typically functions normally, resulting in a disappearance of the seizures and therefore, the epilepsy. On a larger, macroscopic scale, other processes are likely involved. In patients suffering from a head trauma, neuronal (and glial) death and rewiring of surviving neurons may be responsible for the post-traumatic epilepsy that is observed in a significant portion of these patients. Epilepsy is not only studied in patients. In animal models and brain tissue slices, epilepsy can be induced as well, allowing a detailed study of changes in epileptogenesis [15, 16]. In addition to these *in vivo* and *in vitro* models, mathematical and computer modeling has made substantial progress over the past few years. Epilepsy models range from simulations of a few to more than 100-thousand neurons, sometimes with impressive amounts of detail [17]. It is hoped, that in the forthcoming years our insight into the pathophysiology of this disorder will further increase and contribute to improved treatment of patients suffering from this disorder.

4.4.3 Dementia

Dementia relates to a decline in cognitive function, where at least memory is involved. In addition to the loss of particular memory functions, patients lose various other skills, as well, e.g. language and abstract reasoning. The most common form of dementia is Alzheimer's disease (AD). The number of people suffering from AD is large, and increases with age. While in the age group 65-69 years the incidence is about 3-5 per 1000, it rises to more than 50 per 1000 ($> 5\%$) in the older than 90-year age group. Estimates in the literature vary strongly, however, and are also dependent on the exact criteria used. Educational level (>15 years vs <12 years) is probably also associated with the risk of AD.

Patients with Alzheimer's disease find it very difficult to learn or remember ordinary facts, reflecting the disturbance in declarative memory, Eventually, they lose this ability completely. However, their procedural memory remains (nearly) normal: they can still learn and remember how to do things. For instance, a patient may not be able to tell you what he/she has eaten this morning, but is perfectly able to ride a bicycle or to open a door.

The clinical diagnosis of AD is based on the medical history, and the absence of other known causes. For instance, patients who suffer from multiple brain infarctions may also develop a dementia, but this is not AD, but a multi-infarct dementia. MRI or CT-scans of the brain of patients suffering from AD may show atrophy, but this is not a very reliable sign. The EEG can also be a useful tool in the diagnostic work-up of AD. Typically, the EEG will show a slowing of the various rhythms, and a reduction in reactivity. This refers to the modulations in the EEG that occur as a function of closing or opening the eyes. If a healthy person closes his/her eyes, there is an *increase* in the power of the posterior dominant alpha rhythm. The brain reacts strongly to being (visually) exposed to the environment in healthy persons. In patients suffering from AD, this reactivity is lost. This is similar to the clinical observation: patients with AD lose their interaction with their environment.

On autopsy, the brains of patients suffering from AD show a loss of neurons and synapses in the cerebral cortex, and also in certain subcortical regions. Also, amyloid plaques are seen. These are dense deposits of amyloid-beta proteins that are present outside the neurons. Within the neuron, structures consisting of insoluble twisted fibers are present, often called tangles. The details of the pathophysiology are unknown, and include inflammatory processes, cytokines and calcium ion homeostasis. At present, there is unfortunately no cure for AD and treatment is mainly supportive.

4.4.4 Multiple Sclerosis

Multiple sclerosis (MS) is the prototype example of a central demyelinating disease. In patients suffering from MS, myelin in the central nervous system is damaged. As you remember, it is the myelin that realizes a fast conduction velocity of action potentials, and realizes a relatively long electrotonic length. Damage to the myelin, therefore, will result in a slowing down of or even blockage of the transmission of action potentials. MS is the most common cause of neurologic disability in young adults. The lesions of demyelination are histopathologically characteristic of the disease.

MS owes its name to the presence of multiple sclerotic (hardened) lesions in the brain and spinal cord with multiple scars. Clinical manifestations of multiple sclerosis are varied, ranging from sensory disturbances to motor control, and sometimes even cognitive decline. The initial attacks can be transient and mild. Initial symptoms often reported are changes in sensory perception, for instance numbness in a leg or arm. Other symptoms that are often reported are changes in motor control, e.g. difficulty in walking or opening and closing a shirt. In most patients, the initial symptoms occur over a period of hours to days. The neurological deficits that occur in these patients, and may partially restore, are known as "Schubs". The optic tract is also often involved. If that occurs, patients may complain about a loss of vision in one eye. This typically develops in a few hours[10]. This is known as an *optic neuritis*. Similar to the other parts of the brain that can be involved in the demyelinating process, the inflammation of the optic nerve may also be (partially) reversible. In practice, therefore, in most patients vision is restored.

The diagnosis is typically made from a combination of clinical, laboratory and radiological data. In fact, the definite diagnosis cannot be made until there is evidence of at least two anatomically separate demyelinating events occurring at least thirty days apart. Brain examination by MRI (magnetic resonance imaging) can accurately detect the "white matter lesions". An example of an MRI in a 44-year old patient suffering from multiple sclerosis is presented in Figure 4.10. The disease course is quite variable. Four clinical subtypes exist. *Relapsing-remitting* describes the initial course of 85% to 90% of individuals with MS. This subtype is characterized by unpredictable attacks (Schubs or relapses) followed by periods of months to years of relative quiet (remission) with no new signs of disease activity. Deficits suffered during the attacks may either resolve or may be permanent. When deficits always

[10] This time relationship (and the age of the patient) differentiates this form of monocular blindness from ischaemia of the retina due to a thromboembolic process. This latter typically occurs in seconds. In addition, if retinal ischaemia occurs, patients often have the experience that a curtain is closed, starting at the top of the visual field.

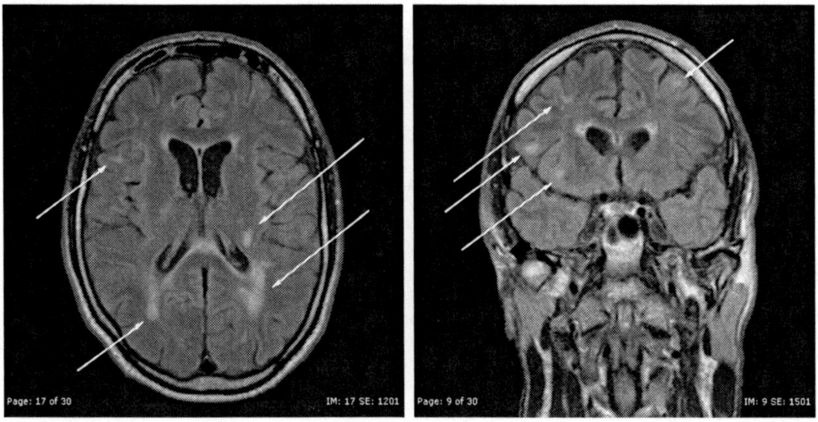

Fig. 4.10 Transversal (left) and coronal (right) magnetic resonance images (T2 flair) of the brain of a 44-year old patient, suffering from multiple sclerosis. Several white matter lesions can be seen (arrows). Images kindly provided by R. Bezooijen, radiologist, Medisch Spectrum Twente, Enschede, Netherlands.

resolve between attacks, this is referred to as "benign" MS. This benign subtype is present in about 10% of the patients. *Secondary progressive* describes around 80% of those with initial relapsing-remitting MS, who then begin to have neurologic decline between their acute attacks without any definite periods of remission. This decline may include new neurological symptoms, worsening cognitive function, or other deficits. Secondary progressive is the most common type of MS and causes the greatest amount of disability. *Primary progressive* describes the approximately 10% of individuals who never have remission after their initial MS symptoms. Decline occurs continuously without clear attacks. The primary progressive subtype tends to affect people who are older at disease onset. Finally, *progressive relapsing* describes those individuals who, from the onset of their MS, have a steady neurologic decline but also suffer superimposed attacks. These four subtypes are important, not only for prognosis, but also for therapeutic decisions.

The precise cause of MS is still unknown, despite tremendous research efforts over the past decennia. It is generally acknowledged that MS is an auto-immune disorder, with often an unpredictable course. On microscopic examination, perivascular inflammation and demyelinisation is observed, eventually leading to sclerotic plaques. T-cells and other immune effector populations that enter the brain "attack" the nerve cells, destroy the myelin and oftentimes destroy the axons and entire remaining structures, as well.

There is no definite cure for MS. Present treatment modifies the progression of the disease, reducing both the number and severity of attacks

("Schubs"). These drugs are known as disease-modifying drugs: they do not cure the disease, but modulate the course.

4.4.5 Stroke

Stroke is a leading cause of mortality and morbidity. A stroke is defined as a sudden loss of neurological function either due to insufficient blood supply to a part of the brain (ischaemic stroke, 80% of cases) or caused by an intracerebral hemorrhage. Insufficient perfusion of brain structures may result from local atherosclerosis, but an important other mechanism is a blockage of (part of) an artery by a thrombo-embolus. In this case, part of an atherosclerotic plaque that is formed in an artery is disrupted, and carried towards a smaller artery, that supplies blood to a part of the brain. Typical locations where these emboli can originate include the carotid artery and the heart.

If neurons are supplied with insufficient oxygen and glucose (oxygen-glucose deprivation (OGD)), the first process that fails is the synaptic transmission [18]: neurons will be compromised in their ability to secrete a neurotransmitter. This will immediately result in a severe loss of neuronal function. Depending on the duration and severity of the OGD, additional energy requiring processes will fail, as well. Ultimately, the sodium-potassium pump will fail, which will result in cell-swelling, and finally, cell death. An MRI of a patient who suffered from a stroke, is shown in Figure 4.11.

The clinical manifestations are varied, and are determined by several factors, including the part of the brain that is affected and the residual perfusion. Strokes that affect those parts of the brain that are supplied by the medial cerebral artery may result in a contralateral paresis or paralysis. If the dominant hemisphere is affected, language can be compromised, as well. In addition, hemihypaesthesia may be present. The long-term outcome of patients who suffered from a stroke is varied, ranging from minimal residual loss of function to complete paralysis and aphasia. As mentioned before, any structural damage to the brain increases the likelihood of developing epilepsy. If this happens in stroke patients, this is called post-stroke epilepsy. This occurs in a significant fraction of patients, up to 7%. Details of the processes that are responsible for post-stroke epilepsy are currently not well understood.

4.4.6 Dysphasia[11]

Language is a very complex process, and many brain regions are involved in the production. These regions are (mainly) located in the dominant hemisphere. More than 140 years ago, Paul Broca in 1864 and Carl Wernicke in 1874, proposed a first classification of aphasias, and a characterization of

[11] Information for this section on dysphasia, in particular the two conversations between the patients B.L. and H.W.,was found on the internet `http://pages.slc.edu/ ebj/IM_97/Lecture10/L10.html`

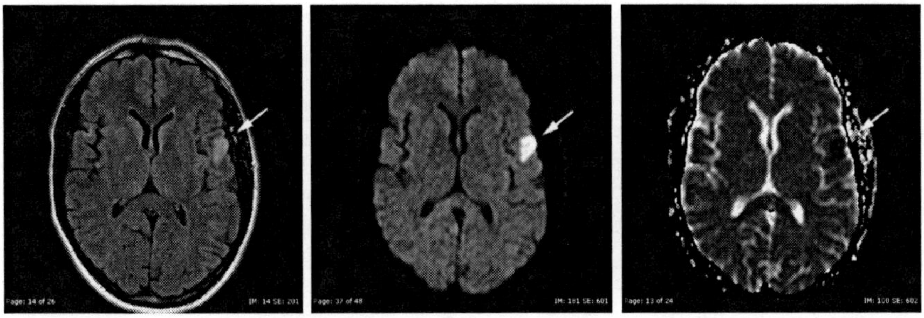

Fig. 4.11 T2-flair (left), diffusion-weighted image (DWI, middle) and apparent diffusion coefficient (ADC, right) MR of a 54-year old patient with acute dysarthria and dysphasia, showing a well circumscribed cortical ischaemic lesion. Since diffusion is low in the infarcted area, the signal intensity is high (white), with a low signal intensity in the ADC-image. Images kindly provided by R. Bezooijen, radiologist, Medisch Spectrum Twente, Enschede, Netherlands.

a simple brain circuit for the processing of language. They obtained their insights from post-mortem studies of patients, who suffered from language disorders. Broca defined the language center located in the frontal hemisphere, and discovered that this center was essential for *producing* language. His first patient, who was also known as "Tan", since this was the only word that he could speak, suffered from syphilis. Post-mortem examination of this brain showed that Tan had a lesion in the left frontal cerebral hemisphere, that Broca held responsible for his language problem. This area is now known as Broca's area, about 2-3 cm in diameter. Soon after, Wernicke published about other brain areas, that -if damaged-, resulted in language disorders, as well. He found that damage to the left posterior, superior, temporal gyrus, results in problems with the comprehension of language. This region is now referred to as Wernicke's area. In almost all right handed people this is the left hemisphere. In about 75% of left-handed people, the left hemisphere is the dominant hemisphere, as well[12]. The two important areas involved in language are shown in Figure 4.12. An acquired difficulty in the use of language or the understanding of language is referred to as an aphasia or dysphasia. This is most often caused by an ischaemic or hemorrhagic stroke, a trauma, or an infection in the dominant hemisphere. Damage to Broca's region, as observed by Broca in his patient "Tan", gives rise to a slow, speech

[12] So, how do we know this? This knowledge is nowadays not only based on post-mortem research, but also on findings obtained with functional MRI. This allows us to perform various experiments to study language lateralization. These studies, e.g. [19], have shown that in almost all right-handed people, the left hemisphere is involved in language processing. In a few percent, however, there is bilateral involvement. In left-handed people, bilateral activation occurs in about 15%, with right hemisphere lateralization in only 10%.

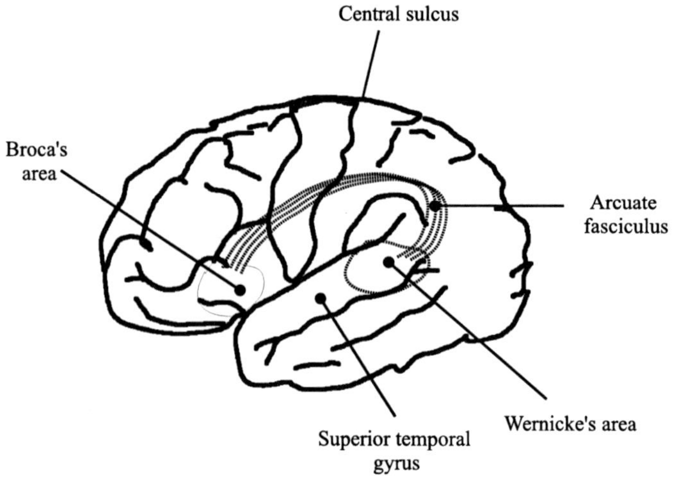

Fig. 4.12 Location of Broca's and Wernicke's area in the left hemisphere.

that lacks ordinary rhythm and intonation. The patient's speech is agrammatical. Words like "to" and "be" are not used while he is speaking. Below is an example of a Broca's aphasic, B.L., describing what is happening in a picture of three people in a kitchen. This picture shows a women doing the dishes. While the woman has forgotten to turn off the water, that is spilling on the floor, her two children are trying to get a cookie. The boy is standing on a chair, almost falling, while his little sister is encouraging him. This is illustrated in Figure 4.13.

Fig. 4.13 The "cookie theft" picture from the Boston Diagnostic Aphasia Examination. Picture reprinted with kind permission of PRO-ED, Inc., TX, USA.

The conversation with the Broca aphasic B.L. now went as follows:

B.L.: Wife is dry dishes. Water down! Oh boy! Okay Awright. Okay ...Cookie is down...fall, and girl, okay, girl...boy...um...
Examiner: What is the boy doing?
B.L.: Cookie is...um...catch
Examiner: Who is getting the cookies?
B.L.: Girl, girl
Examiner: Who is about to fall down?
B.L.: Boy...fall down!

Patients with Wernicke's or fluent aphasia produce a quite different speech. Although the ordinary grammatical structure is intact, the content of the speech is not. A Wernicke's aphasic produces a nonsensical speech, consisting of short sequences. The pattern is rapid, and fluent, with a normal rhythm and intonation. Although various sentences of the Wernicke's aphasic may make sense in isolation, in the context of the other sentences, the language is quite meaningless. The language is often repetitive and broken up by nonsense words or "new" words (neologisms), made up by the patient. An example of such a neologism is *perspecial.* The following is a sample from an aphasic patient (H.W.), asked to describe the same cookie theft picture.

H.W.: First of all this is falling down, just about, and is gonna fall down and they're both getting something to eat...but the trouble is this is gonna let go and they're both gonna fall down...but already then...I can't see well enough but I believe that either she or will have some food that's not good for you and she's to get some for her too...and that you get it and you shouldn't get it there because they shouldn't go up there and get it unless you tell them that they could have it. and so this is falling down and for sure there's one they're going to have for food and, and didn't come out right, the uh, the stuff that's uh, good for, it's not good for you but it, but you love it, um mum mum (smacks lips)...and that so they've...see that, I can't see whether it's in there or not.
Examiner: Yes, that's not real clear. What do you think she's doing?
H.W.: But, oh, I know. She's waiting for this!
Examiner: No, I meant right here with her hand, right where you can't figure out what she's doing with that hand.
H.W.: Oh, I think she's saying I want two or three, I want one, I think, I think so, and so, so she's gonna get this one for sure it's gonna fall down there or whatever, she's gonna get that one and, and there, he's gonna get one himself or more, it all depends with this when they fall down...and when it falls down there's no problem, all they got to do is fix it and go right back up and get some more.

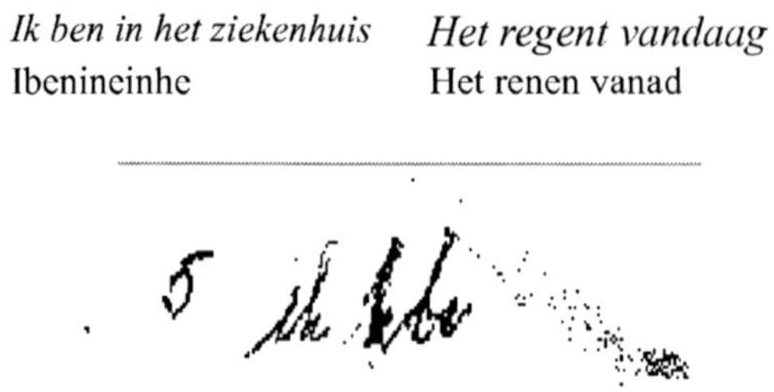

Fig. 4.14 Result of writing the sentence: "Ik ben in het Ziekenhuis" (I am in the hospital) and "het regent vandaag" (It is raining today) using a typewriter. Italics show the correct sentences. At the bottom, the patient uses a pen. This is a case of severe dysgraphia.

Besides damage to the primary language areas, Broca or Wernicke, various language disorders are due to damage to connections between these and other areas. This may result in very remarkable clinical syndromes, for instance a patient may not be able to write, while its capacity to understand language and to produce spoken language is preserved. This condition is known as agraphia; an example is presented in the next subsection.

4.4.7 *Agraphia*

As a final example, we present a patient that was seen a few years ago at our department. This patient had both an impressive and at the same time, subtle complaint. He told us, that he had been experiencing difficulty in the last few weeks in writing or typing words. He was used to compose, every other week, a little booklet, containing text and hymns for the Sunday Mass in his local church. This last weekend, he had found it extremely difficult to put any word on paper using his typewriter. He experienced similar difficulties in writing using a pen. While he was telling about his complaints, only very subtle problems in his language were noticed, where, now and then, there was a minor hesitation. When asked to write down a sentence, he produced the results shown in Figure 4.14. The differential diagnosis included an infarction or a tumor. Since the symptoms did not appear to have started very suddenly, but rather gradually, a brain tumor was the most likely diagnosis. Unfortunately, a malignant brain tumor was indeed diagnosed (astrocytome, grade IV). He was treated with dexamethason and scheduled for neurosurgery. A few days later, his agraphia had disappeared. Unfortunately, he died a few weeks later due to complications from surgery.

4.5 Summary

This chapter discusses some processes related to interneuronal communication and neural circuits. An elementary neuron model, the integrate-and-fire concept, is discussed. You globally understand the concepts of neural plasticity and Hebbian learning. Some circuits relevant for clinical neurology are presented, including several examples of neurological diseases characterized by disturbances in normal "circuit behavior", like epilepsy, multiple sclerosis, stroke, and Parkinson's disease.

Problems

4.1. Compare and contrast the functional characteristics of an electrical synapse and a chemical synapse.

4.2. Synapses are not present on:
a. the cell body
b. the dendrite
c. the axon hillock
d. none of the above

4.3. Outline the process of neurotransmitter release into a synapse. Start with the arriving action potential and end with neurotransmitter release.

4.4. Mention three possibilities to treat Parkinson's disease.

4.5. When a neurotransmitter is said to be "excitatory" what does that mean with regard to: 1) the post-synaptic potential, 2) the ion flux at the post-synaptic membrane, and 3) the probability that the post-synaptic neuron will fire?

4.6. What are the possible fates of a neurotransmitter released into the synaptic cleft?

4.7. Contrast 1) temporal and 2) spatial summation of a post-synaptic potential (PSP).

4.8. What is an EPSP and an IPSP? What effect does each have on the probability that post-synaptic neuron will fire? How are graded potentials involved in this process?

4.9. A specific structure in the synapse is:
a. the synaptic vesicle
b. mitochondria
c. cytoplasm
d. None of the above

4.10. GABA is a neurotransmitter which opens Cl- gates on the post-synaptic membrane. Based on this information, is GABA probably an excitatory or inhibitory neurotransmitter? Explain your answer.

4.11. A right-handed patient with epilepsy experienced a period in which he/she could not speak very well, while understanding what was being asked. After about 10 min the symptoms were gone. These findings indicate
a) an ictal focus around Broca
b) an ictal focus around Wernicke
c) an ictal focus in the right temporal lobe
d) this is known as paroxysmal dysarthria
e) None of the above

4.12. Sketch an RC-network of an electrical synapse. What is the typical transmission time of an electrical synapse?

4.13. Read about the various stretch reflexes. What are the anatomical connections relevant in the knee jerk reflex? And what components are relevant in the evaluation of the biceps and triceps stretch reflex?

4.14. Can you mention some examples of a cutaneous reflex? What is a Babinski sign?

4.15. Mention a few upper motor neuron diseases and give some examples of lower motor neuron disorders.

4.16. List at least 4 differences that are found on clinical examination between an upper and a lower motor neuron disease.

4.17. Using Eq. 4.5, sketch the firing rate of a model neuron using τ_m=30 ms, $V_{\text{rest}} = V_{\text{reset}} = -65$ mV, $V_{\text{th}} = -50$ mV, and R_m=90 MΩ. Take for the injected current, I_e values from 0 to 2 nA. You should find frequencies between 0 and 400 Hz.

4.18. Write your own Matlab program to simulate an integrate-and-fire neuron. You can add some more realism to the model, if you add a noise term to the threshold value.

References

1. Eichenbaum, H.: The cognitive Neuroscience of Memory: An introduction. Oxford University Press, Oxford (2002)
2. Khalilov, I., Holmes, G.L., Ben-Ari, Y.: In vitro formation of a secondary epileptogenic mirror focus by interhippocampal propagation of seizures. Nat. Neurosci. 6(10), 1079–1085 (2003)
3. Izhikevich, E.M.: Dynamical Systems in Neuroscience. MIT Press, Cambridge (2007)
4. Izhikevich, E.M., Edelman, G.M.: Large-scale model of mammalian thalamocortical systems. Proc. Natl. Acad. Sci. U S A 105(9), 3593–3598 (2008)
5. Trappenberg, T.P.: Fundamentals of Computational Neuroscience. Oxford University Press, Oxford (2002)

6. Snell, R.S.: Clinical Neuroanatomy. Lippincott Williams & Wilkins (2006)
7. Fitzgerald, M.J.T., Gruener, G., Mtui, E.: Clnical Neuroanatomy and Neuroscience. Elsevier Saunders, Amsterdam (2007)
8. Catani, M., Ffytche, D.H.: The rises and falls of disconnection syndromes. Brain 128, 224–2239 (2005)
9. McGeer, P.L., McGeer, E.G., Suzuki, J.S.: Aging and extrapyramidal function. Arch. Neurol. 34(1), 33–35 (1977)
10. Carlsson, A., Winblad, B.: Influence of age and time interval between death and autopsy on dopamine and 3-methoxytyramine levels in human basal ganglia. J. Neural Transm. 38(3-4), 271–276 (1976)
11. Bernheimer, H., Birkmayer, W., Hornykiewicz, O., Jellinger, K., Seitelberger, F.: Brain dopamine and the syndromes of parkinson and huntington. clinical, morphological and neurochemical correlations. J. Neurol. Sci. 20(4), 415–455 (1973)
12. Wolters, E.C., van Laar, T., Berendse, H.W. (eds.): Parkinsonism and Related Disorders. VU University Press, Amsterdam (2007)
13. Wasterlain, C.G., Treiman, D.M. (eds.): Status epilepticus: Mechanisms and Management. The MIT Press, Cambridge (2006)
14. van Putten, M.J.A.M.: Neonatal seizure detection. Clin. Neurophysiol. 119(11), 2417–2418 (2008)
15. Schwartzkroin, P.A. (ed.): Epilepsy: models, mechanisms and concepts. Cambridge University Press, Cambridge (1993)
16. Pitkänen, A., Schwartzkroin, P.A., Moshé, S.L. (eds.): Models of Seizures and Epilepsy. Elsevier Academic Press, Amsterdam (2006)
17. Soltesz, I., Staley, K. (eds.): Computational Neuroscience in Epilepsy. Academic Press, London (2008)
18. Bolay, H., Gürsoy-Ozdemir, Y., Sara, Y., Onur, R., Can, A., Dalkara, T.: Persistent defect in transmitter release and synapsin phosphorylation in cerebral cortex after transient moderate ischemic injury. Stroke 33(5), 1369–1375 (2002)
19. Pujol, J., Deus, J., Losilla, J.M., Capdevila, A.: Cerebral lateralization of language in normal left-handed people studied by functional MRI. Neurology 52(5), 1038–1043 (1999)

Part II
Clinical Neurophysiology

Chapter 5
Basic Measurement Principles

> I often say when you can measure what you are speaking about and express it in numbers you know something about it; but when you cannot measure it, when you cannot express it in numbers, your knowledge of it is of a meagre and unsatisfactory kind.
>
> — William Thomson (Lord Kelvin) (1824-1907)

5.1 Introduction

This chapter discusses various basic measurement principles as used in Clinical Neurophysiology to acquire bio-signals. We start with a general treatise on transducers, and present various issues like accuracy, sensitivity, drift and linearity. We also discuss the processes that occur if ionic currents are to be converted into electronic currents.

Since the earliest developments in medicine and biology, physicians and scientists have been interested in the measurement of medical (biological) processes. Many of the first developments in this area were made by researchers that were both educated in the medical field and the basic sciences like physics or mathematics. Examples are Gilbert, Copernicus, Bernoulli, Thomas Young en Helmholtz, who were both physician and physicist. Other well known researchers who contributed significantly to quantification of phenomena in the medical field were Leonardo da Vinci (1452-1519), Galileo Galileï (1564-1642) and the anatomist Luigi Galvani, who discovered "animal electricity" in 1785.

The pulse frequency was measured for the first time by Sanctorius of Padua (1561-1636). This was a far from trivial achievement, since the clocks that were known at that time were not able to measure minutes. Sanctorius used the so called "pulsilogium", that consisted of a lead bullet, attached to a string. If the bullet was put into motion, the frequency of the swings

M.J.A.M. van Putten: Essentials of Neurophysiology, BIOMED, pp. 91–113.
springerlink.com

was a function of the length of the string. This is described by the period $T = 2\pi\sqrt{l/g}$, with l the length of the string and g the gravity acceleration[1]. By comparison of the frequency of this "pendulum" and the number of counts of the pulse, a measure for the pulse frequency was obtained. Today, fortunately, this can be performed much easier, and many transducers are currently available for biomedical signal acquisition, including biosignal recording.

The applications of transducers in medicine are widespread, and reach from a simple scale to measure the weight of an individual to highly advanced measurements of the brain function, for instance the magnetoencephalogram (MEG), measured with SQUIDS[2], intra-cerebral electrodes or functional MRI (fMRI). Depending on the specific application, the sensors and actuators need to be more accurate, reliable and safer. This chapter deals with some general aspects of (medical) transducers, where the main emphasis will be on the sensors.

5.2 Man as Object for Measurement

The human being generates signals from various energy domains, many of which can be measured by invasive or non-invasive techniques. Electrical signals (bio-electricity) are generated by virtually all cells, and are one of the most relevant signals for information transport in the central and peripheral nervous system. Note, that in this process magnetic fields are present, as well. Measurements devoted to measuring these bio-electrical signals are for instance electroencephalography, electromyography and magnetoencephalography.

Chemical information is carried by neurotransmitters and hormones. The human being also produces infrared radiation, which is the basis for thermography. Mechanical energy is generated by the action of muscles, both voluntary and involuntary (e.g. the heart and bowels). On the other hand, we can react to various stimuli, given the fact that we possess particular *sensors*.

As an example, we can react to an intensive mechanical stimulus, leading to tissue damage, by experiencing pain. Typically, this will lead to a kind of withdrawal response. Different tissues have different absorptive characteristics for radiation, which is used in radiology. Table 5.1 present a short overview of parameters that can be measured, including the typical frequency range and sensor characteristics. It is clear from this table, that almost all sensors convert energy to the electrical domain. In this domain, it is easy to transport electrical signals, and to store and process this data.

---

[1] Given that the angle, α, between the string and the vertical, is relatively small.

[2] Super-conducting QUantum Interference DeviceS.

Table 5.1 Various physiological variables used in clinical medicine. Information partially from [1].

parameter	range	frequency	sensor
blood pressure	10 - 400 mm Hg	DC-50 Hz	indirect: Hg manometer direct: manometer
heart sounds	threshold 100 µPa	DC-2 KHz	stethoscope microphone
CVP	0 - 40 mm Hg	DC-30 Hz	catheter
cardiac output	4-25 l/min	DC-20 Hz	thermodilution
blood flow	1-300 ml/s	DC-20 Hz	electromagnetic ultrasonic
pO_2	2-15 kPa	DC	biochemical electrode
pCO_2	0 - 10 kPa	DC	biochemical electrode
pH	6.8-8.0	DC-2 Hz	biochemical electrode
saturation	0-100%	DC	pulse-oximeter
ECG	0.5-4 mV	1-250 Hz	biopotential electrode
EEG	5-300 µV	0.5-35 Hz	biopotential electrode
MEG	0.1-5 pT	0.5-150 Hz	SQUID
MCG	10-50 pT	1-250 Hz	SQUID
action potential	0.01-3 mV	DC-10.000 Hz	biopotential electrode
resp. frequency	2-50 /min	DC-10 Hz	strain gauge
temperature	20-45 °C	DC	thermometer thermistor radiation thermometry
air flow	0-600 l/min	DC-40 Hz	differential pressure sensor hot wire

5.3 Clinical Neurophysiology

As previously discussed, the human organism uses a wealth of signals[3] from different energy domains (chemical, electromagnetic, mechanic, etc) for the control and regulation of several processes and its interaction with the environment. This interaction can be both self-initiated (active modulation of the environment) as passive (receiving), as illustrated in Table 5.2.

Due to many different causes, normal function may be disturbed. Examples include motor neuron diseases, trauma, infections, etc. However, the human being has a limited number of disease *presentations*, and we typically need additional resources to define the exact underlying cause. As an example, a patient may complain about a loss of muscle control. This could be caused by a disturbance of the peripheral motor neuron (motor neuron diseases), but another possibility is a disturbance of the motor control system in the central nervous system. Therefore, clinical examination is often accompanied by investigations such as imaging (MRI, CAT scan), and more function related investigations to come to a specific diagnosis. Examples of these auxiliary

[3] Signals can be defined as forms of energy that carry information.

Table 5.2 Overview of various bio-signals as generated by the organism and input signals that it can process.

Internal and output	input
chemical: neurotransmitters	chemical: taste, smell
electrical: heart, brain, muscles	electrical: stimulation
radiation: warmth	radiation: heat, light
temperature	temperature
mechanical: speech, motor	mechanical: sound, touch

investigations are electroencephalography (EEG), MEG, electromyography (EMG), functional MRI (fMRI), PET and SPECT.

Clinical neurophysiology is one of the various specialties in clinical medicine, that is specialized in using particular measurement techniques to assist in the (neurological) diagnostic process. In addition, clinical neurophysiologists have contributed to an increased understanding of various (neurological) diseases by studying the processes responsible for several neurological diseases, such as various neuropathies, epilepsy, autonomic dysfunction, or myasthenia gravis.

Clinical neurophysiology is involved with the registration and interpretation of various bio-signals as spontaneously generated by the organism. In addition, clinical neurophysiology studies the reaction of the organism to a particular input stimulus. Both processes yield information about the normal and abnormal functioning of various systems. Moreover, we can induce changes in the functioning of various systems by application of particular stimuli. Clearly, some elementary insight into the code or language of the system at hand is needed to do so in a responsible and successful way. Examples include modulation of central circuits in patients suffering from abnormalities in motor behavior, epilepsy or chronic pain. In sum, therefore, clinical neurophysiology involves two aspects of signal analysis:

- Analysis of (spontaneous or induced) bio-signals to increase insight into function of various neurophysiological processes for diagnostic purposes, including monitoring;
- Control of processes by active application of signals. Examples include deep brain stimulation in depression, Parkinson's disease or epilepsy.

The biosignals that are available to us are most often electrical signals, generated by the action of neurons or muscle cells. Clinical neurophysiology also uses different signals than electrical bio-signals, such as mechanical energy from strain gauges as applied in polysomnography to determine the excursions of the thorax during breathing, or a temperature distribution.

5.4 General Aspects of the Measurement Process

The measurement process is the activity with which information is retrieved from the environment in a form that is suitable for a particular interpretation (*e.g.* the measurement of the ambient temperature). Information can loosely be defined as a certain ordering present within an energy domain that reduces the uncertainty about a particular phenomenon. Information transfer is the process with which information is converted within a single energy domain or from one energy domain to another (for instance the storage of an image to CD-ROM). The process of acquiring a signal that is assumed to contain relevant information about a particular system is known as data-acquisition. This is not always easy. For instance, in the beginning of the former century, electronic amplifiers with input impedances $\gg 1\ \text{M}\Omega$ did not exist, nor digital filters or computer displays. Signal acquisition, therefore, was not a trivial process.

5.4.1 Definition of a Transducer

Transducers can be defined as energy converters within an energy domain or between energy domains, where ideally the information contents should not change. Transducers can be found both at the input and at the output of measurement systems. In acquiring data, there is a *sensor* at the input, typically followed by signal processing, and finally, a representation to the end-user, for instance displaying a graph or numbers on a computer screen. This latter part is also know as the *actuator*. Schematically, this process of signal acquisition is shown in Figure 5.1, which is the basic configuration of any measurement system.

An example of a sensor is the electrode that is used to record the ECG. This transducer converts bioelectrical signals (carried by ionic currents) to

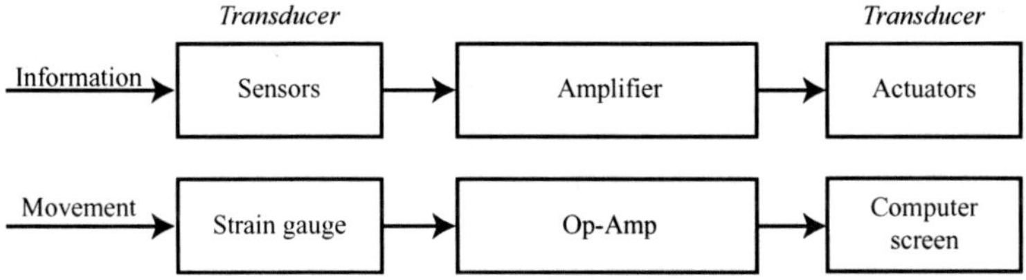

Fig. 5.1 General process of signal acquisition. The transducer at the input is the sensor, the transducer at the output is the actuator; oftentimes, this is a computer screen. In various circumstances, more than one actuator is present for interpretation of the signal, e.g. both a display and a loudspeaker. An example is needle electromyography (discussed in a later chapter), where the physician uses both visual information from the computer screen, displaying the wave forms, and the sonofication of the electrical activity of the muscle under investigation.

electronic currents: i.e. electrical signals carried by electrons. Another example of input transducers are the light sensitive cones and rods in the retina, that convert photons to action potentials, i.e. a conversion from electromagnetic radiation to ionic currents.

An electrode can also act as an actuator, as for instance used in a pacemaker that stimulates the heart muscle[4] or for deep brain stimulation.

Transducers that are neither at the input or at the output of a system have no special names. The process that is performed at this level, does, however, and is called *signal conditioning*, such as filtering or amplifying. In the human body, transducers include particular neuronal networks that may filter the input signal, or reshape it such, that transport over a longer distance is possible.

What is relevant in this discussion, as well, is that transducers *do not add new information to the signal.* They shape, filter, or transport information, but they are not "creative". The brain, however, *can* combine input from various transducers, and process the information, and does seem to add new information to the information present in the sensing input. The brain is considered to be "creative", although the amount of creativity differs tremendously between different organisms, including man. In this respect, therefore, the brain seems to possess self-generating mechanisms for new knowledge.

5.4.2 Modulating and Self-generating Transducers

An important characteristic of transducers is whether they need an auxiliary energy source for their operation. A *modulating* transducer is defined as a transducer that needs an auxiliary energy source for its operation. A well known example of a modulating transducer is a strain gauge. The electrical current that flows through this device is modulated by mechanical energy, in particular force, inducing a change in resistance. The input signal, therefore, modulates the energy of the energy source. Another example of a modulating transducer is an LCD display: you need light for its operation. Alternatively, self-generating transducers are e.g. a solar cell, a thermocouple (input transducers), a heating element or a light bulb, where electrical current is converted into light. Self-generating transducers that do not realize an energy conversion are also know as *shape transducers* or *modifiers*. Examples include a heat exchanger or a gear box.

The human body contains a wealth of transducers, such as touch receptors, pain receptors, receptors for gravity, acceleration, etc. We can view the bones of the inner ear as a modifier or shape convertor, illustrating a biological self-generating transducer. An overview of all these transducers is presented in Table 5.3.

---

[4] Typically, modern pacemakers possess a sensing electrode, as well, often combined in "one" electrode.

Table 5.3 Overview of sensory receptors in the human body.

Type	Stimulus	Location
Touch	pressure	skin
Cold	temperature	skin
Warmth	temperature	skin
Pain	tissue damage	mostly in skin
Auditory	pressure	Inner ears
Taste	chemical	tongue
Olfactory	chemical	nose
Visual	radiation (light)	Retina
Temperature	change from 37°C	brain
Vestibular	acceleration	inner ears
chemoreceptor	chemical	arteries
baroreceptor	pressure	arteries
kinesthetic	angle	joints
spindle	stretch	skeletal muscles
tendon organ	mechanical	tendons

5.5 General Characteristics of Sensors

The value x that is assigned to the measurement of the parameter X is called the measurement result or *measurand*. Ideally, the sensor measures exactly the true value, x_t, of the process under consideration. Many influences exist, however, that may cause the value of the measurand x to deviate from the true value x_t. This section will describe the contribution of the several phenomena that cause this deviation and the terminology used to describe these deviations.

5.5.1 Systematic and Non-systematic Errors

The difference between the measurand, x and the true value, x_t, is known as the total (measurement) error or inaccuracy. This error can subsequently be divided into two causes: *systematic* errors and *non-systematic* errors, where it holds that the total error is the sum of these two contributions.

Non-systematic errors influence the measurand in an unpredictable way, while systematic errors are constant, both in size and (if applicable) direction. Non-systematic errors can in principle be totally eliminated by repeating the measurement, yielding the *mean* of the measurand, $\overline{x}$, defined as

$$\overline{x} = \frac{1}{N}\sum_{i=1}^{N} x_i \tag{5.1}$$

where N is the total number of measurements performed. If N is sufficiently large, it can be proven that the mean, $\overline{x}$, will in the limit be equal to the expected value of the measurand, μ. The expected value of $\overline{x}$ equals therefore μ. Put in another way, $\overline{x}$ is a non-biased estimator for μ; in formula $E[\overline{x}] = \mu$. This does *not* imply, however, that the estimate is equal to the true measurement value, x_t, since there may still be a contribution from systematic errors. A systematic error is also known as *bias*. An example is the measurement of body weight, with a scale that has an off-set of e.g. 3 kg. This will cause all measurements to have a systematic error (or bias) of +3 kg.

An estimator, therefore, can be biased yielding an inaccurate estimate. If the influence of the systematic errors can be neglected, the measurement is accurate, and we have $E[\overline{x}] = x_t$.

5.5.2 Precision

A measure of the contribution of the random or non-systematic error is called the precision (or sometimes imprecision). The precision, therefore, is a measure for the spread of the single estimates around the expected value, μ, in repeating the measurement. If the precision is high, the measurement is called precise. If the measurement has a numerical character, which is very often the case, the precision is the number of significant numbers of the measurement. The value 135 mV is therefore less precise than 135.32 mV. Another notation for the precision of a normally distributed measurement is the standard deviation, for instance 120.3 ± 0.4, where the standard deviation $\sigma = 0.4$. A high precision does *not* imply that the measurement is accurate, however, as will be discussed next.

5.5.3 Accuracy

The accuracy of a single measurement is the number that describes the total measurement error, *i.e.* both the non-systematic and the systematic errors. A possible notation is the deviation, $x - x_t$. More often, however, the accuracy is presented relative to the true value, as

$$\frac{\text{total error}}{\text{true value}} = \frac{x - x_t}{x_t}. \tag{5.2}$$

The accuracy, therefore, is a measure for the total measurement error, independent of the contribution of this error. The concepts accuracy, precision and bias are finally illustrated in fig. 5.2.

5.5.4 Resolution

The smallest incremental quantity that can be measured with certainty is the *resolution*. Can you estimate the difference in resolution of an 8-bit and a 12 bit AD convertor, if scaled in the range 0-1 V?

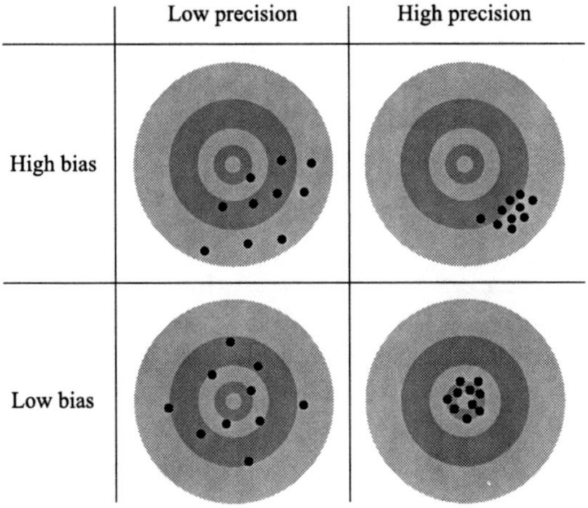

Fig. 5.2 Illustration of precision, bias and accuracy using a dart board. The low bias low precision situation can arise if the darter drunk too much beer. The high bias, high precision situation is possible if the darter is not wearing his spectacles, which may result in a systematic error. The champion realizes both low bias and high precision. The author is aware that a real dart board looks slightly different.

5.5.5 *Sensitivity and Drift*

Ideally, sensors should sense only the defined physical input, S, for which they were developed. A physical input can be, for example, electromagnetic radiation, mechanical or chemical energy. In real world applications, however, a sensor is sensitive to quantities other than S as well. Mathematically this can be expressed as

$$U_{\text{out}} = \gamma U_{\text{in}}(S) + E(U_1, U_2, \cdots, U_n) \tag{5.3}$$

where U_{out} is the output signal of the sensor, $U_{\text{in}}(S)$ the function that relates S to the output signal, and $(U_1, U_2, \cdots, U_n)$ any other undesired physical input signals that affect U_{out}[5].

The value of γ is the *sensitivity* of the sensor, defined for constant E, for which it holds therefore that

$$\gamma = \frac{\partial U_{\text{out}}}{\partial U_{\text{in}}}. \tag{5.4}$$

In general it further holds for γ that

$$\gamma = \gamma(U_{\text{in}}, U_1, U_2, \cdots, U_n). \tag{5.5}$$

The sensitivity, therefore, can in principle be a function of the input signal.

[5] The effect of $(U_1, U_2, \cdots, U_n)$ on the output signal is also known as cross-sensitivity.

All variations in γ are defined as *multiplicative drift* or *sensitivity drift*, while a variation in E, due to changes in $(U_1, U_2, \cdots, U_n)$, is known as *additive drift*, or *zero drift*. Drift can arise by changes in supply voltage, non-linearities, changes in temperature, mechanical vibrations, etc.

We wish to realize a situation given by

$$U_{\text{out}} = \gamma U_{\text{in}} + \epsilon(U_1, U_2, \cdots, U_n) \tag{5.6}$$

with γ is *constant*. Sensors that satisfy Eq. 5.6 are ideal: the output signal is mainly a function of the measurand, U_{in}, and other physical parameters influence the output signal with a maximal contribution of ϵ, only.

Many different techniques are applied in trying to realize the situation, as described by Eq. 5.6, as for instance discussed in [3, 4].

5.5.6 Linearity

Linear systems possess the important property of superposition: the response to the sum of a input signals is the sum of the output signals that was obtained with the individual input signals. Using y_1 as the response to the input signal x_1 and y_2 as the response to the input signal x_2, a system is linear if and only if

1. the response to $x_1 + x_2$ is $y_1 + y_2$;
2. the response to ax_1 is ay_1, where a is any (complex) constant.

Linear systems have the important (and pleasant) characteristic that the behavior for any input signal can be described by a single function, that is independent of the input signal.

5.6 Dynamic Characteristics of Transducers

The characteristics of transducers discussed so far relate to static (or DC) input signals. Most parameters that are obtained in the biological or medical measurement processes are not static, but dynamic in character, however, and change as a function of *e.g.* time or position. Examples are the time varying electrical signals of the heart or the brain and the levels of some hormones in the blood serum. In order to satisfactory describe the relationship between the input and output signal of sensors that measure these qualities or actuators that generate a time varying output signal, differential equations are needed. Since, on the time scale of the measurement process, the characteristics of the transducer will often not change[6], and many transducers are (approximately) linear[7] over the range of interest, it is often possible to apply the theory of linear differential equations with constant coefficients as a suitable first order description of their behavior.

[6] This is not true in general, for instance chemical sensors may have characteristics that are a function of the exposure time to (a constant) concentration of a substance.

[7] The transducers are either linear by design or linearized.

5.7 Converting Ionic to Electronic Currents: Interfaces

Electrical sensors are needed for the conversion of bio-electricity, generated by the various ionic currents in our body, into *electronic* currents, and are also known as biopotential electrodes. There are many applications of these sensors, especially in cardiology, neurology and intensive care medicine. We will discuss the main measurement principles of the sensors used in the recording of the ECG and the EEG. In a later section, we will present some actual examples of currently used electrodes in a particular application.

In the measurement of a biological electrical process, which is based on the flow of ions, the ionic currents need to be converted into electronic currents. These electronic currents are subsequently transported to our actual measurement instrument, e.g. a volt meter or oscilloscope. The interface function between the body currents and the electronic currents is carried out by biopotential electrodes. During a particular measurement of biopotentials, some current will flow in our electrical circuit to make the measurement possible. It is desired that this current is as small as possible in order to allow a reliable measurement of the process at hand[8].

Metal (or metal plated) electrodes are used in the measurement of these ionic currents. At this electrode-skin interface, the conversion of ionic to electronic currents takes place. In general, two different electrode designs are possible. The electrode can behave like a resistor, and actual charge transfer across the interface is possible. The electrodes that behave as such are known as non-polarizable electrodes. On the other hand, the electrode can behave like a capacitor, and no actual charge is transferred across the interface, and the electric current is a *displacement* current. Of course, the frequency characteristics of these two electrodes are different: the non-polarizable electrode can measure up to DC-voltages, while the polarizable electrode cannot, and behaves more like a high-pass filter. In the real-world, neither of these electrodes can be fabricated, and actual electrodes behave more or less like either polarizable or non-polarizable electrodes.

5.7.1 Polarizable Electrodes

Ideal polarizable electrodes do not allow charge transfer across their interface, and measure body potentials by allowing displacement currents to be induced in the electrode material. Electrodes that are made of a noble metal (silver, gold, platina) come closest to the perfectly polarizable electrodes. In clinical EMG measurements, the concentric needle electrode can be coated with platinum, and then behaves like a polarizable electrode. Given the nature of the biosignals that is explored (see also Chapter 9 on electromyography), the high-pass filtering characteristics of these electrodes pose no problem, since the relevant frequencies are typically well above 50 to 100 Hz.

[8] This is typically realized by using high input-impedance amplifiers. For EEG signals, for instance, this is typically larger than 10 MΩ.

5.7.2 Non-polarizable Electrodes

The processes that take place the interface of non-polarizable electrodes are quite complex. We will discuss various aspects in some more detail below[9].

Let us start our discussion by looking at the phenomena that occur if two conducting substances are placed into contact, for instance a metal and an electrolyte. In these circumstances, potential differences can arise across the interface. This voltage difference is known as the *half-cell potential* in the case of a metal-electrolyte boundary[10]. As we will shortly show, the expression for this potential difference is basically the Nernst equation.

Let us put a metal, made of atoms M, in a solute, containing one of his ionic solutions. For instance, let us take Aluminum (Al) and put it into a glass of drinking water (that contains Aluminum salts). At the metal boundary, the following reactions will take place:

$$Al \rightleftharpoons Al^{3+} + 3e^{-}. \tag{5.7}$$

Therefore, at the metal boundary, charges will be separated, as illustrated in Figure 5.3. This charge separation results in a voltage difference, the electrostatic potential, also known as the half-cell potential. The electrolyte contains anions A^{-}, that may react according to

$$A^{-} \rightleftharpoons A + e^{-} \tag{5.8}$$

assuming a monovalent anion.

The electrode therefore can become oxidized, hereby producing a cation and one or more electrons. The cation will diffuse into the electrolyte, while the electrons remain in the electrode. Anions from the electrolyte may also become oxidized, releasing one or more electrons and creating a neutral atom. Besides oxidation, reduction reactions will occur, where reactions in the reverse direction take place. If no current flows through the electrolyte-metal interface, there is equilibrium between the oxidative and reductive reactions, where the nett electron current is zero. If there is a current flowing through the interface, it depends on the direction which reaction will play the major role. If there is a current from the electrode to the electrolyte, the oxidation reactions will dominate; in the reverse direction the reduction reactions will be the most significant. We will not further discuss the details of the processes that take place in this layer; we do mention, however, that movement

[9] Some parts in this section were strongly inspired by the (much more extensive) treatment on biopotential electrodes in Medical Instrumentation, Applications and Design, John G. Webster (editor), John Wiley & Sons, 1998 [1].

[10] You can do these experiments at home, if you like. Simply take a glass of tap water, and immerse an aluminum plate. Connect one probe of your voltmeter to the aluminum, put the second probe straight into the water. You will now measure a potential difference. This potential difference is caused by the differences in half-cell potentials between the water-aluminum interface and the water-probe interface, and is typically around 500 mV. You can repeat the experiment using different metals.

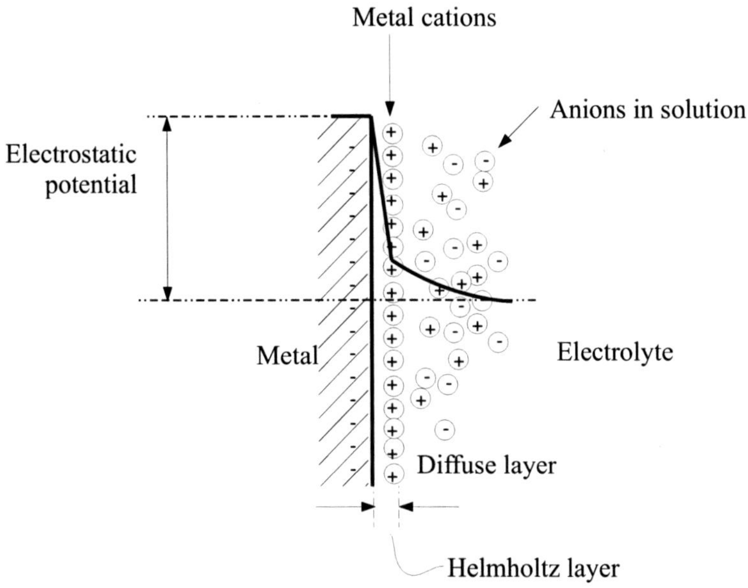

Fig. 5.3 Metal-electrolyte interface. At the metal-electrolyte interface, there is a slight separation of charge (cf the cell membrane), the amount of which is a function of the concentration of the metal salts in the solution and the temperature. Note the Helmholtz double layer, that separates the electrons remaining in the metal from the metal cations, that are now in solution. This layer is very thin, of molecular dimensions (typically $3 - 4 \times 10^{-10}$m).

of the metal-electrolyte interface will affect the double layer, and may induce temporarily disturbances in the equilibrium. This may lead to voltage differences, and is one of the causes of *movement artifacts.*

The height and stability of the voltage difference across the metal electrolyte interface depends amongst others on the valence of the metal ions, the ease with which ions can go into a solution, the concentration of the ions in the electrolyte, and the presence or absence of current flow through the interface. The *standard* half-cell potentials are measured relative to a hydrogen electrode with, by definition, a half-cell potential of zero volts.

The standard half-cell potential is defined and determined at standard temperature and in an electrolyte with cations of the electrode material with unit activity[11]. In most situations, the (actual) half-cell potentials deviates

[11] The activity of an ion can be regarded (in this context) as an "effective concentration". It is a dimensionless constant, defined as $a_i = \gamma_i C_i / C_\theta$, with $0 < \gamma < 1$. C_i is the concentration (in one of its units) and C_θ an arbitrary, *non-zero* standard (deliberately chosen to be so). At low concentrations, the activity of a solute can be approximated to the ratio of its concentration over the standard concentration, and is therefore approximately equal to its concentration.

Table 5.4 Standard half-cell potentials for some common electrode materials at 25°C. By definition, the half-cell potential of hydrogen is zero.

Metal and reaction	**potential E^0 (Volt)**
$Al \rightarrow Al^{3+} + 3e^-$	-1.706
$Zn \rightarrow Zn^{2+} + 2e^-$	-0.763
$H_2 \rightarrow 2\,H^+ + 2e^-$	0.000
$Ag + Cl^- \rightarrow AgCl + e^-$	+0.223
$Ag \rightarrow Ag^+ + e^-$	+0.799
$Cu \rightarrow Cu^{2+} + 2e^-$	+0.340
$Cu \rightarrow Cu^{1+} + e^-$	+0.522

from the standard half-cell due to differences in temperature and activity, according to

$$E = E^0 + \frac{RT}{F}\ln(a_{c^{n+}}) \tag{5.9}$$

where

E = half-cell potential
E^0 = standard half-cell potential
n = valence of the electrode material
$a_{c^{n+}}$ = activity of the cation C^{n+} (mol/l)
R = the universal gas constant
T = the absolute temperature
F = Faraday's constant

Polarization

The half-cell potentials that arise at the metal-electrolyte interface are in general not constant, but a function of *e.g.* temperature, the kind of metal that is used for the electrode, and the presence or absence of current flow through the interface. The change in the half-cell potential due to this current flow is known as the *overpotential*, which is caused by polarization of the electrode. Of course, this is in general an undesirable phenomenon, especially if the overpotential is a function of the current in the desired measurement range.

Three mechanisms are known to contribute to the arisal of overpotentials: Ohmic effects, concentration changes and activity changes. All these three contributions relate to the required energy to convert an ionic current to an electronic current at the metal-electrolyte interface.

The *ohmic overpotential* arises due to dissipation of energy to overcome the resistance of the electrolyte.

The *concentration overpotential* may arise by changes in ion concentration in the electrolyte at the metal-electrolyte interface due to the passage of current. This can be pictured as follows. If no current is present in the electrolyte, an equilibrium will exist between the metal ions in the electrolyte and the metal atoms in the electrode and the anions in the electrolyte and the electrons in the electrode. If current passes, this no-current-equilibrium will shift towards a new equilibrium situation, with a change in half-cell potential, due only to changes in the concentration of the ions that take part in the reactions.

Finally, the *activation overpotential* is a factor of influence in the generation of the overpotential. The oxidation and reduction reactions at the electrolyte-metal interface are not completely reversible in the sense that there exists a difference in activation energy between the oxidation and reduction reactions. If current passes, either the oxidation or reduction reaction will dominate. This difference in activation energy will manifest itself in the occurrence of a voltage drop, which is known as the activation overpotential.

These three mechanisms are additive, so we have

$$V_p = V_r + V_c + V_a \tag{5.10}$$

with

V_p = total overpotential or polarization potential
V_r = ohmic overpotential
V_c = concentration overpotential
V_a = activation overpotential

In the design of various recording electrodes, the goal is to develop electrodes with a polarization potential, V_p, as small as possible. To realize this goal, all three possible contributions to the polarization potential can be reduced, as will be discussed further.

Electrodes that show no polarization effects at all are called *perfect non-polarizable* electrodes. With these electrodes, the energy barrier for the conversion of ionic to electronic currents, with the passage of an electron through the metal-electrolyte interface, is zero. At the other end of the spectrum, there is the class of perfect *polarizable* electrodes, that we discussed earlier, where the energy needed for an electron to pass the boundary is infinite. A perfect polarizable electrode therefore behaves as an ideal capacitor. Electrodes made of a noble metal approximate polarizable electrodes; the silver-silver chloride electrode approximates the qualities of a non-polarizable electrode and is widely used. This electrode will be more extensively discussed in the next section.

Besides the choice of the kind of metal for the electrode, other possibilities to reduce the influence of the overpotential is the use of amplifiers with a high input impedance, to minimize the currents (in amperes per square meters) through the metal-electrolyte interface and the use of large electrode areas for the same reason.

The silver-silver chloride electrode

The silver-silver chloride electrode is the best example of a very common electrode that approximates the perfect non-polarizable electrode. The electrode consists of a silver basis, covered with silver-chloride. The electrolyte that is being used contains chloride ions, and is usually saturated with silver-chloride, in order to prevent solution of the surface of the electrode. At the interface, the following reactions will take place.

$$\begin{cases} Ag \rightleftharpoons Ag^+ + e^- \\ Ag^+ + Cl^- \rightleftharpoons AgCl \end{cases} \tag{5.11}$$

where the second reaction occurs immediately after the oxidation of the silver atoms. For the overall reaction, we can write

$$Ag + Cl^- \rightleftharpoons AgCl + e^- . \tag{5.12}$$

This silver chloride is badly soluble and will precipitate at the surface of the electrode. This realizes a near constant Ag^+ concentration. At the electrode where the electrons are "generated", oxidation takes place; at the other electrode, electrons are "gained": reduction. Using the solubility product, K_s, given by

$$K_s = a_{Ag^+} \cdot a_{Cl^-} \tag{5.13}$$

it holds for the half-cell potential, according to Eqn. 5.9, that

$$E = E^0_{Ag} + \frac{RT}{F} \ln \left(\frac{K_s}{a_{Cl^-}} \right) \tag{5.14}$$

which can be rewritten as

$$E = E^0_{Ag} + \frac{RT}{F} \ln(K_s) - \frac{RT}{F} \ln(a_{Cl^-}). \tag{5.15}$$

Since K_s is constant for a given temperature, it holds that the half-cell potential is a function of the temperature and the activity of the chloride ions. Since this last concentration is nearly constant, realized by using an electrolyte with a high concentration of chloride ions, this realizes the near perfect non-polarizable electrode. The processes are schematically illustrated in Figure 5.4.

Although the Ag/AgCl electrodes come near the perfect non-polarizable electrode, there still remain (although relatively constant) non-zero voltage differences across the electrode-skin interface, due to remaining polarization effects. This would pose no problem, if these voltage differences would be exactly the same for all the electrodes used in a particular measurement. This is, however, not the reality: the electrode-skin voltages will vary between different Ag/AgCl electrodes, resulting in non-zero voltage differences, even if no signal is present. These voltages are typically larger than the biosignals

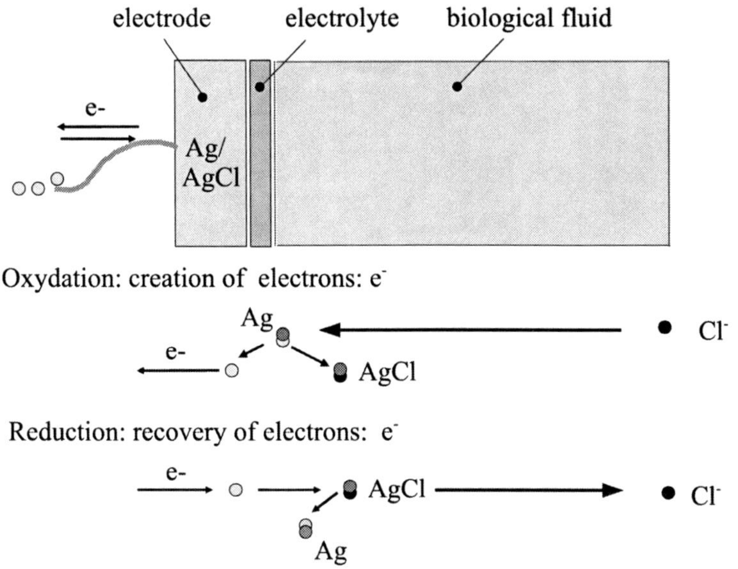

Fig. 5.4 Schematic of the redox reactions that take place when using silver-silver chloride electrodes. The electrolyte solution contains Cl^- ions as the principle anion, and should be saturated with AgCl. Chloride anions are also present in the biological fluid.

we like to measure. EEG signals, for instance, are of the order of 10-100 µV, only. Superposed on this DC-component, we will now record our signals of interest. Fortunately, however, many relevant biosignal have frequencies above 0.5 Hz. Therefore, we can use an analog high-pass filter to eliminate this DC-component, for instance with a high-pass setting of 0.5 or 1 Hz (as is also indicated in Table 5.1).

5.8 Electrical Analog of the Electrode-Skin Interface

As we discussed before, the ideal electrode is non-polarizable and therefore shows no over-potential. These electrodes, therefore, have current and frequency *independent* properties. In practice, it is difficult to completely realize electrodes with these properties. Therefore, the current-voltage relationship across the metal-electrolyte interface is often a nonlinear function of both the current and the frequency. Not only at the electrolyte-metal boundary, but also at the electrolyte-skin interface processes take place that may disturb the measurement process. The equivalent circuit to model the complete skin-electrolyte-electrode interface is illustrated in Figure 5.5.

An important phenomenon that plays a role in the measurement process is the movement artifact: the occurrence of voltage changes due to the movement of the electrode. These voltage changes mainly occur at the

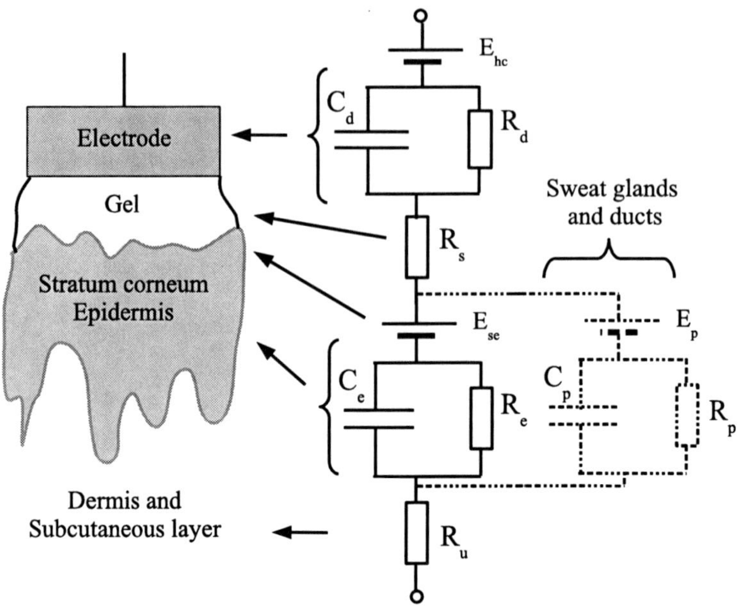

Fig. 5.5 Equivalent circuit of the electrode-electrolyte-skin interface. The electrode resistance and capacitance are R_d and C_d, respectively. These make up the impedance associated with the electrode-electrolyte interface and polarization effects. The electrode half-cell potential is E_{hc}. $\mathrm{R_s}$ is the gel resistance and $\mathrm{E_{se}}$ the voltage difference that can arise at the gel-epidermis (at least stratum corneum) interface. $\mathrm{R_e}$ and $\mathrm{C_e}$ represent the impedance of the epidermis. R_u is the resistance of the dermis and subcutis. The parallel C_pR_p combination in series with E_p reflects contributions of the sweet glands and ducts. These components are often neglected, but are relevant in the galvanic skin reflex. Redrawn from Medical Instrumentation, Application and Design, John W. Webster, editor, John Wiley & Sons, New York, © 1998.

electrolyte-skin interface and the electrode-electrolyte interface. Using sand paper, to carefully abrase the upper skin layer (the *stratum corneum*) the first effect can be reduced. The voltage changes at the electrode-electrolyte interface are mainly due to changes of the double layer and occur most significantly with the use of polarizable electrodes. In silver-silver chloride electrodes this contribution is smaller, however not always negligible.

5.9 Biopotential Electrodes

A large number of biopotential electrodes to measure electrical phenomena exist. The maximal allowable size of the electrode depends on the specific application. For elementary monitoring of the heart rate, the size will not be critical. For the measurement of very local activity, where a high spatial

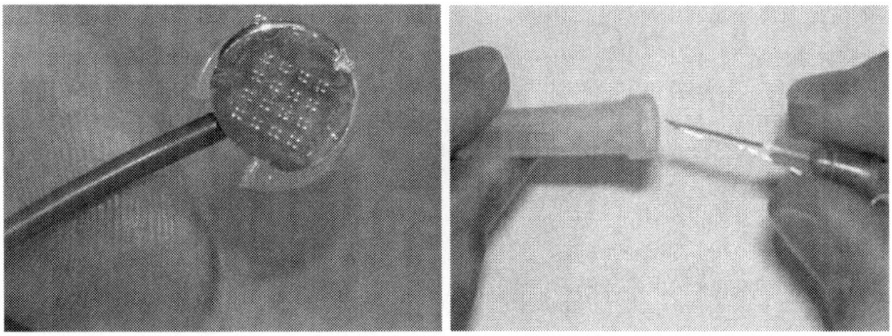

Fig. 5.6 Left: The microneedle array electrode, with 64 microneedles (courtesy of Dr Luttge, MESA+, University of Twente, Enschede, Netherlands). Since this electrode pierces the stratum corneum, it belongs to the class of minimal invasive electrodes. Right: Wire electrode with the needle that is used for insertion of the wire. This is an example of an invasive EEG electrode, designed for intensive care unit monitoring [5].

resolution is required, very small electrodes are needed. In addition to the division of electrodes into polarizable and nonpolarizable types, electrodes are characterized by their need to cross the skin barrier in the actual applications.

5.9.1 Noninvasive Electrodes

Noninvasive or surface electrodes are used to measure biopotential phenomena at the skin surface, leaving the skin completely intact.The most common noninvasive electrodes are the metal electrode, the suction-electrode and the self-adhesive electrode. The self-adhesive electrodes are usually disposable, and already contain an electrolyte gel. These electrodes are used for instance for the measurement for the electrocardiogram (ECG) or the electromyogram (EMG).

5.9.2 Minimal Invasive Electrodes

Minimal invasive electrodes need some skin preparation in order to allow recording of the biosignal[12]. This may include abrasion of the skin, to remove a part of the stratum corneum. The classical example is the silver-silver electrode, that we previously discussed in detail in section 5.7.2. These electrodes are often used for recording of the electroencephalogram (EEG).

[12] We realize that this definition is somewhat arbitrarily. For instance, skin preparation may involve removing a part of *dead skin cells* only, as is typically needed in the preparation for EEG recording. An alternative definition is that minimal invasive measurements involve removal of or insertion into *living* cells, but without a total penetration of the skin.

In Figure 5.6-left, an experimental microneedle array electrode is shown, that contains 64 microneedles, each about 300 micrometer in length and with a diameter of 100 micrometer. This electrode can pierce the stratum corneum, and is currently evaluated for EEG recordings. An advantage of this electrode, is that additional preparation of the skin is not needed, saving significantly on mounting time.

5.9.3 Invasive Electrodes

Invasive or depth electrodes are electrodes that completely pierce the skin. An advantage of these electrodes is that phenomena that arise at the skin-electrode interface, like movement artifacts, are less likely to occur. Of course, other challenges remain, including biocompatibility. The function of the electrolyte is realized by the body fluid, that contains, amongst others, as pointed out before, a large concentration of chloride ions. Usually, depth electrodes are made of steal or silver (as a coating). They can further be divided into

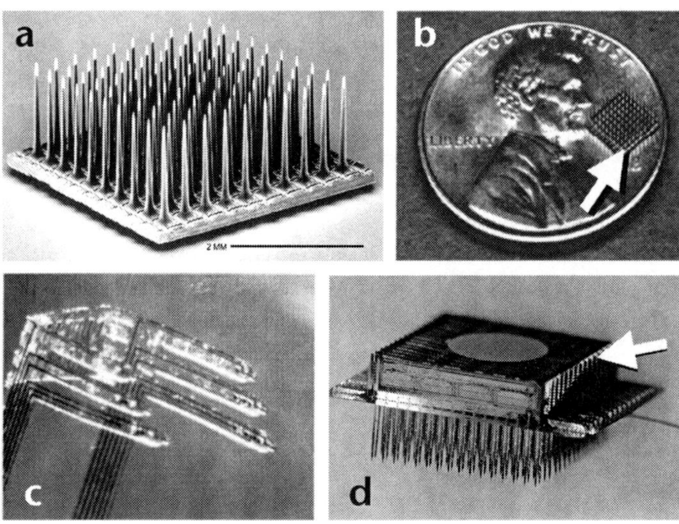

Fig. 5.7 Examples of intracortical electrode arrays under development. (a) Silicon 100 electrode array; each is separated by 400 µm (courtesy of B. Hatt, Cyberkinetics, Inc.). (b) Silicon array shown against a common scale to illustrate the size of these devices (courtesy of E. Maynard, University of Utah). (c) Polyamide "bioactive electrode" array (courtesy of D. Kipke, University of Michigan). (d) Michigan thin film 256-shank array of 1024 multiplexed sites with mounted signal processing electronics (arrow; courtesy of K. Wise, University of Michigan). Illustration republished from John P Donoghue, Connecting cortex to machines: recent advances in brain interfaces, *Nat Neurosci*, 2002, 5 Suppl, 1085–1088, Nature Publishing Group, England.

internal and *percutaneous* electrodes. The internal electrodes have no direct physical contact with the outside world, and are for instance applied in a cardiac pacemaker. A common percutaneous electrode is the needle electrode which is used for the intramuscular EMG recording. Another example is the percutaneous wire electrode, that can be used for continuous EEG recordings. This wire electrode is inserted using a small needle. An example of this electrode is shown in Figure 5.6-right.

Glass electrodes with a tip diameter of typically tenths of µm can be used for patch-clamping, allowing the measurement of very small ionic currents through one or a few ion channels [6].

Modern microtechnological examples of intracortical electrode arrays are shown in Figure 5.7. These electrodes can be used for brain-machine interfacing, as a means to aid paralyzed patients in communication [7]. We will discuss brain computer interfaces to some more extent in Chapter 10.

5.10 Summary

Transducers at the input of a measurement system are sensors, at the output actuators. Sensors can be characterized by both static and dynamical characteristics. You understand concepts as accuracy, inaccuracy, imprecision, resolution, bias, sensitivity and drift. You understand various of the processes that occur in measuring biosignals, including the conversion of ionic to electronic currents. You know the difference between polarizable and nonpolarizable electrodes. In addition, you can give some examples of noninvasive, minimally invasive and invasive electrodes. A very good reference for further reading is [1].

Problems

5.1. What is the difference between a modulating and self-generating transducer? Give examples of three transducers from each group.

5.2. What is the relation between sensitivity drift and additive drift to the systematic and non-systematic errors?

5.3. Mention some sensors with their possible zero-drift and sensitivity drift influences.

5.4. Using two different blood pressure meters, A en B, the diastolic blood pressure is measured, as indicated in Table 5.5. As a reference, another meter is used, that has no systematic error.
a. Estimate the real blood pressure, x_g, of this patient, using the true mean calculated from the standard, $\mu = \overline{x}$. What are the mean values and the standard deviations obtained with meter A and B, respectively?

Table 5.5 Blood pressure as measured with meter A and B. The standard meter is the reference.

standard	80	83	88	90	91
A	160	175	190	200	210
B	70	73	78	80	81

b. What are the systematic errors in measurements A and B. Which meter is most precise? What is the sensitivity, K, of meter A and B? Is K independent of the input signal? (Make a graph of K versus blood pressure) You should realize that the blood pressure changes significantly over a day, with variations up to 60 mm Hg. It is therefore more sensible to define an interval, in between the blood pressure will be in 95% of the times that the measurement is performed. Usually, this is assumed implicitly. Explicitly, the diastolic blood pressure of this patient will be for instance 82-86 mm Hg.

5.5. A thermometer is made of a circular reservoir with radius R and a cylindrical capillary, C, with a diameter of 0.2 mm. Assume that the room temperature is about 20 °C, and that the volume of the reservoir, V_R, is much larger than the column of the capillary, $V_R \gg V_C$.

What is the minimal value of the radius, r, of the reservoir V_R to ensure that the sensitivity of the thermometer is at least 2 mm per degree Celsius? The cubic expansion coefficient of mercury is $0.182\ 10^{-3}\ K^{-1}$.

5.6. A person holds an electrode with a surface of 1 cm^2 in each hand. His skin has the skin resistances at 50 Hz as shown in the table below. If the skin is completely pierced, a resistance of 200 Ω remains. Calculate for each of the three situations depicted in the table the maximal allowable voltage if the current density is not allowed to exceed 5 mA. Calculate these values, too, if the skin is pierced (for instance with a needle electrode).

Table 5.6 Skin resistance at 50 Hz in three different situations with an electrode surface of 10 cm^2.

dry skin	100 kΩ
wet skin	50 kΩ
with electrode gel	10 kΩ

5.7. What is the difference between the half cell potential and the overpotential? Mention some causes for the overpotential. What possibilities exist to reduce the overpotential. What is an ideal electrode?

5.8. Calculate the half cell potential for a Ag-AgCl electrode. Write the expression as a constant plus the natural logarithm of the chloride concentration. The solubility product, $K_s = a_{Ag^+} \times a_{Cl^-} = 1.78 \times 10^{-10}$ at 25°C.

5.9. What is the difference between polarization and the overpotential?

5.10. What measures can be taken to reduce polarization and movement artifacts in the recording of the EEG?

5.11. What is the effect of the removal of the layer of AgCl from a silver-silver chloride electrode?

5.12. You perform a measurement with an amplifier with input impedance of 5 kΩ. You measure a voltage of 5 mV. You perform a second measurement with an amplifier with input resistance of 50 kΩ. Now, you measure a voltage of 25 mV. What is the source impedance?

References

1. Webster, J.G. (ed.): Medical Instrumentation, Application and Design. John Wiley & Sons, New York (1998)
2. Fisch, C.: Centennial of the string galvanometer and the electrocardiogram. J. Am. Coll. Cardiol. 36(6), 1737–1745 (2000)
3. van Putten, A.F.P. (ed.): Electronic Meaurement Systems. Institute of Physics Publishing (1996)
4. van Putten, M.J.A.M., van Putten, M.H.P.M.: Facing drift: a comparison of three methods. Sensors and Actuators A 90(3), 172–180 (2001)
5. Ives, J.R.: New chronic EEG electrode for critical/intensive care unit monitoring. J. Clin. Neurophysiol. 22(2), 119–123 (2005)
6. Ravesloot, J.H., van Putten, M.J., Jalink, K., Ypey, D.L.: Analysis of decaying unitary currents in on-cell patches of cells with a high membrane resistance. Am. J. Physiol. 266(3 Pt 1), C853–C869 (1994)
7. Donoghue, J.P.: Connecting cortex to machines: recent advances in brain interfaces. Nat. Neurosci. 5(suppl.), 1085–1088 (2002)

Chapter 6
Stimulation of Nervous Tissue

While the former chapter discussed various aspects related to performing measurements to *record* bio-signals, this chapter presents various issues related to the *stimulation* of neural tissue, in particular peripheral nerve fibers and brain tissue. In many neurophysiological measurements the system is perturbed or activated with a particular input stimulus, often electromagnetic, and the response is recorded. These measurements are typically performed for diagnostic purposes, e.g. to determine if the velocity of an action potential in a peripheral nerve is sufficiently large. In addition to stimulation of nervous tissue for diagnostic purposes, tissue can also be stimulated for *therapeutic* purposes. If this is used for peripheral nerve or muscle stimulation, it is sometimes referred to as Functional Electrical Stimulation (FES). In the case of stimulation of the central nervous system, as e.g. sometimes applied for the treatment of severe Parkinson's disease or the treatment of some forms of epilepsy [1], the technique is known as (central nervous system) neurostimulation, neuromodulation or deep brain stimulation (DBS). Often, these terms are used interchangeably. In this chapter we will discuss some common elements of stimulation of excitable tissue. The next chapter will discuss in more detail particular clinical applications to evaluate various parts parts of the peripheral nervous system, a field that is known as electro(neuro-)myography.

6.1 Stimulation of Muscle or Nervous Tissue

The notion that many processes in our bodies are controlled by electrical stimuli was discovered in 1783 in Italy. At that time, Luigi Galvani (1737-1798), so the story goes, dissected a frog at a table where he had also been conducting experiments with static electricity. At the moment that Galvani touched the frogs leg with a scalpel, he observed that the leg was moving. It is believed that the scalpel picked up some charge from the electricity machine, that was at the table as well, which led to this discovery of "animal

M.J.A.M. van Putten: Essentials of Neurophysiology, BIOMED, pp. 115–131.
springerlink.com

electricity". Galvani used this term to describe whatever it was that activated the muscles of the specimen. We (and you) know now that most likely voltage gated channels in the peripheral nerve innervating the frogs leg muscle were activated by the charge carried by the scalpel, that led to the observed contraction. Therefore, if muscle fibers or nervous tissue are exposed to an electric field sufficiently large, voltage-gated channels will become activated, and action potentials are generated.

In physiological conditions, nervous tissue is stimulated or activated by the input stimulus it was "designed" for. In fact, this is the actual function of the sensory nervous system: to convey a particular form of energy into the electro-chemical domain to allow further processing by the nervous system. As we learned previously, an obvious example includes the stimulation of the rods and cones in the retina by electromagnetic radiation. Depending on the particular "sensor", the input stimulus is converted into a train of action potentials.

For diagnostic or therapeutic purposes, however, there is a need to stimulate motor or sensory nerve fibers locally, often in non-physiological positions, as well. This is possible by inducing a local depolarization, sufficiently high to exceed the excitation threshold. If this is realized, an action potential will be generated that travels both in the *orthodromic* and the *antidromic* direction, as illustrated in Figure 6.1. This action potential is caused by the activation of the voltage-gated sodium channels present in the cell-membrane, as discussed in previous chapters. Conduction will now occur in two directions:

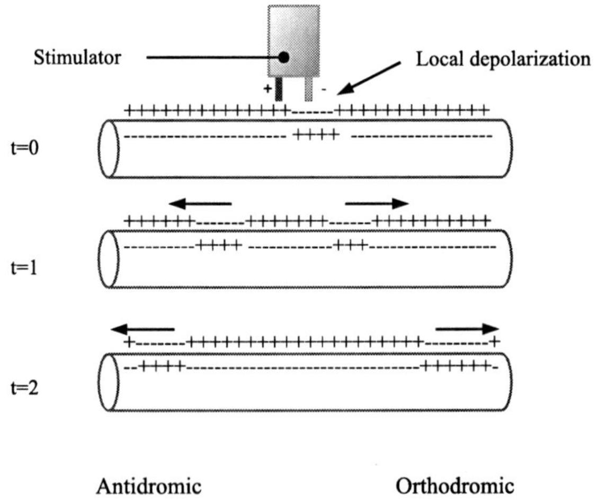

Fig. 6.1 Stimulation of a nerve fiber; antidromic and orthodromic propagation is indicated by the arrows, assuming that the physiological action potential propagation direction is to the right (orthodromic). Note, that the depolarization takes places (approximately) beneath the negative stimulation electrode.

orthodromic conduction in the physiological direction, i.e. towards the target organ, antidromic propagation in the opposite direction. For sensory fibers, orthodromic is towards the brain, for motor nerves it is from the brain, towards the muscle or secondary motor neuron.

6.1.1 Electric Fields in Volume Conductors

The electrodes used for stimulation of neural tissue in *vivo* generate electrical fields. The spatial distribution of these fields is a function of the electrode geometry used and the tissue characteristics. Analysis of the current that flows from these electrodes is important to understand the effects of electrical fields on excitable tissue[1].

The electrodes used for nervous tissue stimulation generally fall into two categories: electrodes that generate a (more or less) uniform field (large electrodes) and electrodes that generate a more localized field (monopolar or bipolar electrodes). To generate a localized field, a constant current or constant voltage source is connected to a small tip of a monopolar or bipolar electrode. To generate a uniform field, large parallel electrodes are used. For homogeneous tissue, it is straightforward to calculate the current distribution given the electric field. For non-homogenous tissue this can be much more complex, in particular due to the anisotropic characteristics. Here, we will only discuss some elementary aspects of the volume conduction in homogeneous tissue. For more extensive treatments, see e.g. [2]. Using the quasi-static formulation of the Maxwell equations, which is allowed at the frequencies considered (< 10 kHz), we can write for the voltage V generated by a single monopolar electrode generating a current I, in a homogeneous volume conductor with conductivity σ

$$V = \frac{1}{4\pi\sigma}\frac{I}{r}. \tag{6.1}$$

Here, r is the distance from the recording point. The electrical field $E = -\nabla V$ and the current density J are now given by

$$E = \frac{1}{4\pi\sigma}\frac{I}{r^2}\mathbf{r} \tag{6.2}$$

and

$$J = \sigma E\mathbf{r} \tag{6.3}$$

with $\mathbf{r}$ the unit radial vector. The electrical field generated by a monopolar electrode, therefore, decays as $1/r^2$. With these electrodes, it is possible to stimulate neurons at more or less *localized* positions. Electrodes used for deep brain stimulation in Parkinson's disease are of this kind.

[1] In the following treatment, an important source of inspiration was the book "Deep Brain Stimulation and Epilepsy", edited by Hans O. Lüders [1], in particular the chapters 4-6.

Relatively large electrodes generate a more uniform field. The voltage $V(x)$ at distance x from the anode by these electrodes, with a voltage V_0 between the electrodes, separated by a distance d, is given by

$$V(x) = V_0 \frac{x}{d} \tag{6.4}$$

where the electric field is given by

$$E = \frac{V_0}{d}. \tag{6.5}$$

Remember, that this kind of electrodes was e.g. used by Hodgkin and Huxley in their voltage clamp experiments, to apply a uniform field across the cell membrane of the squid giant axon, as we also discussed in chapter 2.

When the electrical fields are applied in the neighborhood of neurons and axons, currents will flow, that may enter and exit the neurons. If currents remove charges from the membrane, the membrane potential will decrease, leading to a depolarization. If currents enter the cell membrane, the charge on the membrane capacitance is increased, and the membrane will become hyperpolarized. This is illustrated in Figure 6.2. If sufficient charge is removed by the depolarizing current, voltage gated channels may open up, leading to the generation of an action potential. For simple electrode configurations and homogeneous media, the transmembrane current can be calculated by deriving the effect of the field analytically. For more complex situations, compartmental analysis is needed, e.g. using the method of finite elements. As we discussed in chapter 3, we can model the electrical properties of an axon with the cable equation, that describes the membrane potential as a function of time and place. We previously derived the cable equation, given by

$$\tau_m \frac{\partial v}{\partial t} = \lambda^2 \frac{\partial^2 v}{\partial x^2} - v + r_m i_e \tag{6.6}$$

with the external current i_e. In this derivation, we assumed that the imposed transmembrane current was exactly known, which is a fair assumption if the stimulating electrode is just outside the fiber. However, in most realistic situations, this is not the case, and the stimulating electrode is farther away. In those situations, although it is known what the applied voltage or imposed current is, the externally induced transmembrane current i_e is not known, and must be derived from the partial differential equations that describe the membrane voltage as a function of time, position, external field and membrane properties (membrane and axial resistance and membrane capacitance). In general, when a nerve or muscle fiber is stimulated by an external source, currents will both flow in the extracellular and the intracellular (axial) space, that are of equal magnitude as dictated by Kirchoff's law.

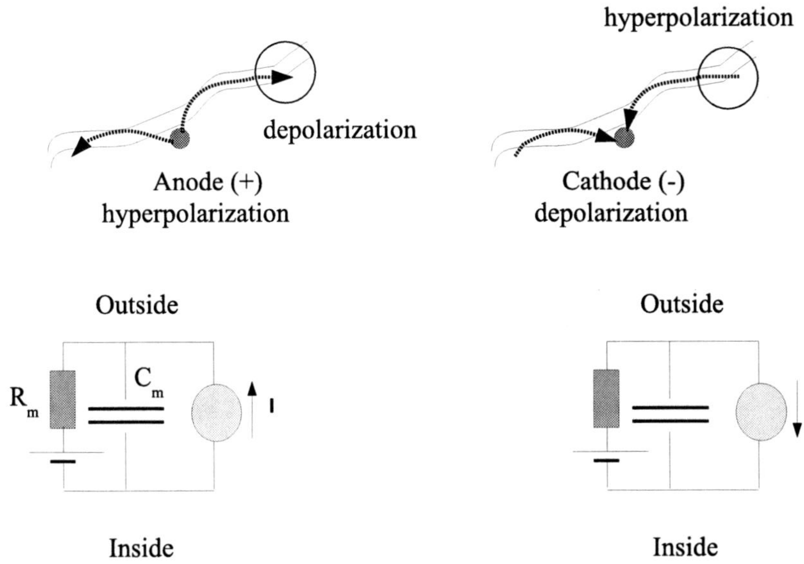

Fig. 6.2 Illustration of membrane polarization and depolarization due to an external current, applied by a monopolar electrode. If current enters the cell, the membrane is hyperpolarized, if current leaves the cell, the membrane becomes depolarized. By convention, the current direction is the direction of the positive charges. Illustration modified and redrawn from Figure 6.1, in Hans O. Lüders (editor), Deep Brain Stimulation and Epilepsy, Martin Dunitz, an imprint of the Taylor & Francis Group, 2004 [1].

A very common situation is to examine the membrane response of a single fiber to the field induced by such a source, located at a particular distance away from the fiber. Depending on the characteristics of the source (e.g. a point source or a uniform field), the field of the source can be described by Eqs. 6.2 or 6.5. Depending on the position of the fiber in the field, transmembrane currents can be induced, resulting in a de- or hyperpolarization. Since the field generated by the fiber itself is generally very small in comparison to the imposed field, in many circumstances the perturbation of the external field by the extracellular field resulting from the presence of the nerve fiber itself can be ignored, especially when the nerve fiber is small compared to its distance from the point source. In many real-world applications and assuming small perturbations, this condition is satisfied. Clearly, the use of such an unmodified field is an important simplification for the evaluation of the excitation of tissue. We will not discuss the various issues involved in deriving expressions for various potential geometries. However, we will present how to modify the cable equation if an external field induced by an external electrode is applied to the nerve fiber. We will now slightly revise the cable equation, based on the electrical circuit shown previously in Figure 3.1, to include the effect of an externally applied electrical field, E_{ext}.

Let us first review the derivation of the cable equation. Introducing the axial current density, I_{ax}, with units ($\text{Cs}^{-1}\text{m}^{-2}$), we can write for the axial current density inside the fiber, I_{ax}

$$I_{ax} r_{ax} = -\frac{\partial v}{\partial x} \tag{6.7}$$

with r_{ax} the specific axial resistance (Ωm). This is simply Ohm's law, stating that the product of the axial current density and the specific axial resistance equals the negative gradient of the membrane voltage. We will proceed with an expression that relates the total membrane current to the axial current. Recall that the total membrane current (per unit length) is the sum of the capacitative and the resistive current,

$$i_m = i_c + i_r. \tag{6.8}$$

For a segment of length Δx we can now write for the total membrane current I_m (Cs^{-1})

$$I_m = 2\pi a \Delta x c_m \frac{\partial v}{\partial t} + 2\pi a \Delta x i_r. \tag{6.9}$$

This current must equal the change in axial current along the fiber, therefore:

$$2\pi a \Delta x c_m \frac{\partial v}{\partial t} + 2\pi a \Delta x i_r = (\pi a^2 I_{ax})_{\text{left}} - (\pi a^2 I_{ax})_{\text{right}}. \tag{6.10}$$

Dividing by $2\pi a \Delta x$ and letting $\Delta x \to 0$ results in

$$c_m \frac{\partial v}{\partial t} + v/r_m = \frac{a}{2} \frac{\partial I_{ax}}{\partial x}. \tag{6.11}$$

Using the expression for the axial current density, Eq. 6.7, we obtain

$$\lambda^2 \frac{\partial^2 v}{\partial x^2} - v = \tau_m \frac{\partial v}{\partial t} \tag{6.12}$$

which is our familiar cable equation. Compare these steps to the derivation resulting in Eq. 3.4.

To obtain an expression for the cable equation that includes the effect of an applied external field, we proceed as follows. For the electrical field inside the fiber, we have

$$E_i = -\frac{\partial v}{\partial x}. \tag{6.13}$$

We will now add the extra (external) field, $E_{\text{ext}}(x, t)$. The total electrical field parallel to the fiber will now equal

$$E_i = -\frac{\partial v}{\partial x} + E_{\text{ext}}(x,t). \tag{6.14}$$

The axial current, I_{ax}, is thus modified, and satisfies

$$r_{ax} I_{ax} = -\frac{\partial v}{\partial x} + E_{\text{ext}}(x,t). \tag{6.15}$$

Again combining the relevant equations, we obtain the modified cable equation

$$\lambda^2 \frac{\partial^2 v}{\partial x^2} - v = \tau_m \frac{\partial v}{\partial t} + \lambda^2 \frac{\partial E_{\text{ext}}(x,t)}{\partial x}. \tag{6.16}$$

This leads to an important general result, that states that the membrane voltage is affected by an external field only if the external field induced by the source has a non-zero second order spatial derivative in the direction along the fiber. This implies that a membrane depolarization or hyperpolarization will not occur if a uniform field ($\partial E/\partial x = 0$) is applied along an ideal cylindrical cell. This is not so difficult to understand. Realize, that it is the *membrane* current that is relevant for inducing a depolarization, not the *axial* current. The membrane current has its maximal value if the spatial gradient of the *axial* current is maximal (cf Eq. 6.7). These gradients in axial currents can be induced by gradients in the electrical field in the axial direction. To stimulate nervous tissue, therefore, spatial gradients of the electrical field in the axial direction of nerve fibers are needed. Real-world neurons can, however, often be stimulated by uniform applied electric fields, as well. This is due to membrane branches or bends: at these points, axial gradients are non-zero, and membrane currents will be induced, that, if sufficiently large, will produce excitation or inhibition. For more details see for instance [1, 2, 3, 4].

6.1.2 Thresholds for Excitation

A point that we will now shortly address is the different sensitivities of neurons for an external field (or current) as a function of their diameter. For instance, C-fibers have a higher stimulation threshold than the larger, myelinated fibers. And myelinated fibers on their turn have lower thresholds if their diameter increases. Realize that currents will enter and leave the axon at those positions where the transmembrane resistance is low, i.e at the nodes of Ranvier in the case of myelinated fibers. It further holds, that the membrane properties at the nodes are relatively constant for all fibers, i.e. the membrane resistance and membrane capacitance at these positions do not significantly differ. Recall, as we discussed in the former section, that it is the second difference of the extracellular potential (or the first difference of the extracellular electric field) in the axial direction that drives the membrane polarization. If we now place a nerve fiber some distance apart from a current source, the currents will now preferentially leave and enter the axon at internodal regions, where the membrane resistance is lowest. The gradients in the axial currents will now be a function of the internodal spacing: if this distance is small, the transmembrane current will be divided by many relatively closely spaced internodes[2], resulting in small transmembrane currents (per unit area) and corresponding small axial current gradients. This implies that relatively large currents are needed to bring this nerve fiber to its firing threshold. Larger internodal spacings will force that the current gradients in the axial direction are much larger, resulting in a lower excitation threshold. Since internodal spacing of myelinated fibers increases (linearly) with external fiber diameter ($l/d_0 \sim 100$), larger diameter fibers have a lower threshold for excitation than smaller diameter fibers. The relationship between threshold current, I_{th}, outside fiber diameter d_o is given by

$$I_{\text{th}} = I_d + \frac{a}{\sqrt{d_0}}. \tag{6.17}$$

where I_d and a are constants[3]. Clinically, these selective sensitivities for an external electrical field are highly relevant. For instance, if motor stimulation is desired, stimulation of sensory nerve fibers should be avoided. In real-world applications, therefore, electrodes are typically placed as close as possible to the nerves that one wants to stimulate. The possibilities for selective stimulation are limited, however. For instance, in vagus nerve stimulation for the treatment of epilepsy, an electrode cuff is positioned around the vagus nerve. This is a mixed nerve, however, containing both sensory and motor fibers. For the dimensions of the electrode cuff, stimulation will typically not only recruit afferent fibers towards the brain, but, due their relatively low stimulation threshold, myelinated motor fibers will be stimulated as well. These motor fibers run to the vocal cords, which may induce hoarseness during the times the stimulator is on. This is one of the undesired side effects in vagus nerve stimulation. Fortunately, they do not occur in all patients, and may even be transient in others.

6.1.3 Strength-Duration Curve

In electrical stimulation of nervous tissue, not only the fiber diameter determines the stimulus strength, but also the duration of the stimulus. This observation that the excitation of nerve fibers is a function of both stimulus strength and stimulus duration dates back more than $\sim$ 100 years. Lapicque (1866-1952), a French physiologist, was one of the pioneers who performed many experiments to study the excitability of nerves. At that time, a central question was: How much and how long a stimulation is required to excite a nerve? Nowadays, this is still a relevant issue, for instance to optimize the settings for deep brain stimulation. The curve that shows the the relation between stimulus strength and stimulus duration to excite a particular nerve

---

[2] Stated differently: there is a high density of low resistance pathways.

[3] More details can be found in e.g. [1].

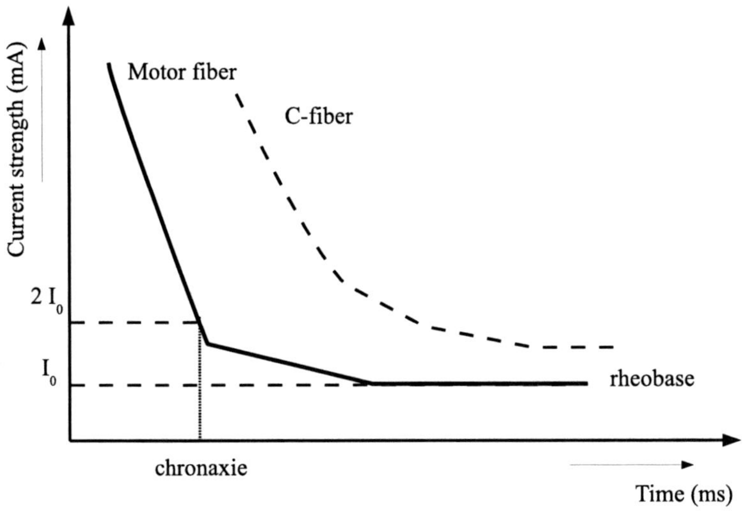

Fig. 6.3 Strength-duration curves for two human nerves. The rheobase current and chronaxie are indicated.

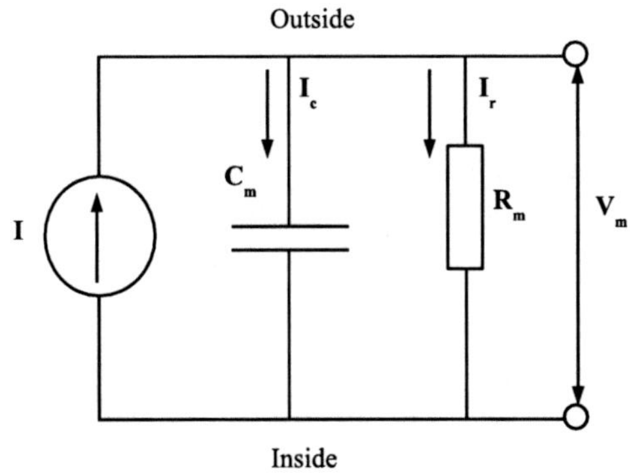

Fig. 6.4 Electrical circuit, representing part of the passive cell membrane. The membrane resistance is modeled as R_m, the membrane capacitance by C_m. The extracellular potential is assumed to be zero.

fiber is the *strength-duration curve*. This curve shows the stimulus intensity needed to achieve a particular response (often the threshold) as function of stimulus duration [2, 5, 6]. An example of such a strength-duration curve is displayed in Figure 6.3. The strength-duration properties are further characterized by the rheobase and the chronaxie. The rheobase is defined as the minimum current intensity needed to obtain excitation with a stimulus of infinite duration. The chronaxie (or time constant, τ_{sd}), is defined as the stimulus duration needed to obtain excitation with a current intensity twice the rheobase. As you can appreciate from Figure 6.3, selective stimulation can be achieved, since the two strength-duration curves are above one another. The nerve fiber belonging to the lower curve can be stimulated if the stimulus strength is below the threshold for excitation of the upper curve. Therefore, in this case, the motor fiber can be stimulated without activating the C-fiber.

To explain the global shape of the strength-duration curve[4], we introduce an elementary electrical circuit, representing a passive cable to model a part of the axon membrane. This elementary circuit is shown in Figure 6.4. The change in voltage that is measured, after starting the current at $t = 0$, with strength I, is described by

$$V(t) = IR_m(1 - \exp^{-t/(R_m C_m)}) \tag{6.19}$$

with R_m the membrane resistance, C_m the membrane capacitance and $R_m C_m$ the RC-time of the circuit. Note, that this is a very simplified model, not incorporating active components, like voltage (and time) dependent sodium and potassium channels. However, it is a useful model for an elementary understanding of the shape of the curves, displayed in Figure 6.3. We will now assume, that the nerve fiber will be stimulated (and generate an action potential) if a particular change in membrane voltage, ΔV, is induced. Rewriting Eq. 6.19, we have

$$I(t) = \frac{\Delta V/R_m}{1 - \exp^{-t/(R_m C_m)}} \tag{6.20}$$

that describes the current-duration relationship. Since the rheobase current is defined as the minimal current to reach the threshold, using an infinite pulse duration ($t \to \infty$), it follows directly from Eq. 6.20 that the rheobase current $I_0 = \Delta V/R_m$. We can now rewrite Eq. 6.20 to

$$I(t) = \frac{I_0}{1 - \exp^{-t/(R_m C_m)}}. \tag{6.21}$$

For a current with duration τ (the chronaxie), it is straightforward to calculate that $t = \ln(2) \cdot \tau \sim 0.693 R_m C_m$. Typical values for human peripheral nerve are rheobase values between 0.5 - 4 mA for peripheral nerve and for cortex. Chronaxie values range from 0.2-0.3 ms for peripheral nerve to 0.4-0.5 ms for cortex. So stimulation of the brain requires larger currents and

---

[4] The equation originally proposed by Lapicque, that describes various experimental data fairly well is given by

$$I = I_0 \sqrt{\frac{t + \theta + \sqrt{(t-\theta)^2 + 0.16\theta^2}}{2t}} \tag{6.18}$$

where I_0 is the rheobase (mA) and θ is 3.8 times the chronaxie (ms). He stated that strength-duration curves of each tissue could be suitably scaled to coincide with a fixed curve, that was therefore identical for all tissues, and called by Lapicque "canonical".

longer pulse durations than stimulation of the peripheral nervous system. The differences that exist in the strength-duration curves of different nerve fibers may be employed for selective stimulation. Analysis of rheobase and chronaxie values is also employed to evaluate diseases of peripheral nerves. Conditions that change the membrane time constant (either by affecting the membrane capacitance and or the resistance) can thus be evaluated. In practice, transcutaneous stimulation of peripheral nerves is by applying a short current pulse on the skin overlying the nerve. Pulse durations range from 0.1-1 ms with currents between 0-100 mA. Bipolar stimulation is the standard. This can induce strong non-uniform electrical fields along the axon and, as you will recall from the former section, this axial gradient is essential to induce changes in the membrane current. If threshold is now reached action potentials are generated and the response to stimulation can be measured by measuring potential differences farther down or up the axon bundle. Additional stimulation at another position than the first stimulus, and measuring the distance between the two stimulation positions, allows estimation of the velocity from the ratio between the distance and the time difference between the corresponding responses. This is discussed in more detail in the next chapter on electro(neuro-)myography.

6.2 Neuromagnetic Stimulation

Besides direct electrical stimulation, nervous tissue can also be stimulated by inducing an electrical field by electromagnetic induction. As you recall, a strongly varying magnetic field will induce an electrical field E, since it holds from Maxwell's equations that

$$E = -\frac{\partial A}{\partial t} - \nabla V \tag{6.22}$$

with A the magnetic vector potential and V the scalar potential. This technique is known as Transcranial Magnetic Stimulation (TMS). One of the advantages above direct electrical stimulation are that this form of stimulation is less painful. The current durations employed in neuromagnetic stimulation are of the order of 0.2 ms. These currents are fed into a typically planar, magnetic coil, with either a round or figure-of-eight design. The electrical field induced will, if the conductivity within a volume conductor (e.g. the brain) is sufficient, induce an electrical current in the tissue. This current, on its turn, can, if large enough, depolarize various neurons in a particular brain region.

This technique was introduced in clinical medicine around 1985 by Barker. He showed that by application of a magnetic pulse above the head, it was possible to induce a current in the brain tissue, that can depolarize cortical neurons. By depolarizing motor neurons, it is possible to induce a movement of a limb, for instance the leg. In this way, non-invasive and relatively painless stimulation of brain tissue is possible. Using this approach it is for instance possible to estimate the central motor conduction time (CMCT) and central conduction velocity. At present, TMS is also increasingly evaluated as a possible therapy for some forms of depression, epilepsy and perhaps even motor neuron diseases. For these therapeutic applications, not a single pulse, but a particular series of pulses is applied: theta burst stimulation. This is a special form of repetitive TMS, that leads, if applied to the motor cortex, to after effects on the corticospinal and cortico-cortical excitability. Most likely, this reflects a form of long-term potentiation and depression. The use of TMS is an intriguing and promising research area, with most likely very interesting clinical applications still ahead.

6.3 Evoked Potentials

An evoked potential is a response or "waveform" that is obtained after repeated application of a particular stimulus, e.g. sound. Since the response to an individual stimulus is typically very small, time-averaging of individual responses is needed to obtain a sufficient signal-to-noise ratio. Of course, only those responses that have a fixed time relationship to the trigger will thus be recorded. The input stimulus can be a visual input (visual evoked potential), auditory (auditory evoked potential) or somatosensory (somatosensory evoked potential). We will briefly illustrate each of these techniques.

6.3.1 Visual Evoked Potentials

If an adequate stimulus is applied to the retina, i.e. visible light, it is possible to record an evoked potential from the occipital lobe. Since the brain continuously generates brain waves, a single stimulus will not be sufficient to realize a signal that can be measured from the "background activity". However, by repeated application of the stimulus, and averaging the stimulus locked responses, the signal to noise ratio can be increased significantly, which may result in a detectable Visual Evoked Potential or VEP.

These VEPs are highly sensitive to lesions of the optic nerve, but relatively insensitive to ophthalmologic disorders. This motivates their use in detecting potential lesions of the optic nerve. VEPs are usually recorded using a checkerboard reversal pattern: the patient is shown black and white squares (on a display), that switch continuously, thus preserving the total illuminance. The strong illumination contrasts between the various black and white squares are a sensitive input signal to the visual system, and result in a time-locked response, that can be recorded over the occipital area. In normal subjects, the visual stimulus evokes a waveform approximately 100 ms after application of the stimulus, the P100. The name relates to the positivity of this waveform, that is recorded about 100 ms after the stimulus. The most common abnormality of the VEP is an increase in this latency. An example of how a normal VEP may look like is shown in Figure 6.5. VEPs may be useful

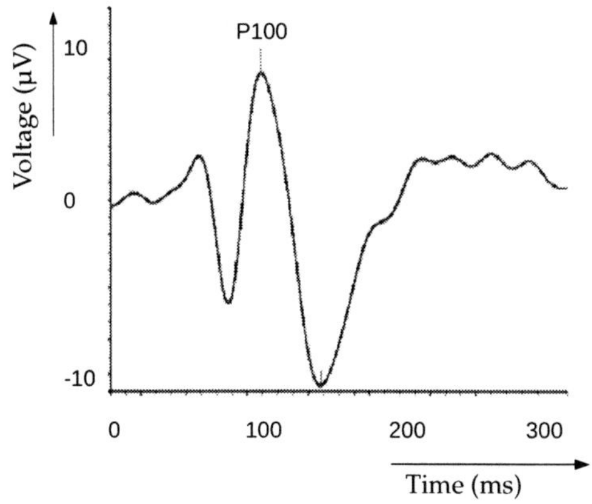

Fig. 6.5 Illustration of a normal VEP.

to demonstrate a lesion of the optic nerve in patients with suspected multiple sclerosis (MS). In many patients with MS, the VEP latency is abnormal, even if no clinical signs are present. The most common VEP abnormality in patients with optic neuritis is a delayed response. This increased latency is caused by the demyelinisation of the optic nerve, resulting in a significant reduction of the conduction velocity, as we discussed in chapter 3.

6.3.2 Somatosensory Evoked Potentials

Somatosensory evoked potentials (SEPs or sometimes abbreviated as SSEPs) are presynaptic and postsynaptic responses recorded over the limbs, spine and scalp following the stimulation of a peripheral nerve. The main value of SEPs is to provide information about the conduction velocity in peripheral nerves, the spinal cord and the brain.

Somatosensory evoked potentials can be induced by different input stimuli, including tendon stretch, stimulation of touch receptors by finger tapping, and electrical stimulation. In most cases, electrical stimulation of peripheral nerves is used, for instance the median, tibial or peroneal nerve.

By averaging a sufficient number of responses (50 to 200), scalp potentials can be recorded with latencies between 20-100 ms, depending on the recording distance and the conduction velocity. Stimulation of the median nerve, for instance, will induce a negative scalp potential after about 20 ms (the N20), that is a reflection of postsynaptic potentials generated by neurons in the hand area of the primary somatosensory cortex in response to the afferent thalamocortical input. Stimulation of the tibial nerve will induce a positive scalp potential after about 38 ms (the P38). Various clinical indications for SEPs exist, including potential spinal cord dysfunction or the localization of sensory symptoms. Patients with conversion reactions or ma-

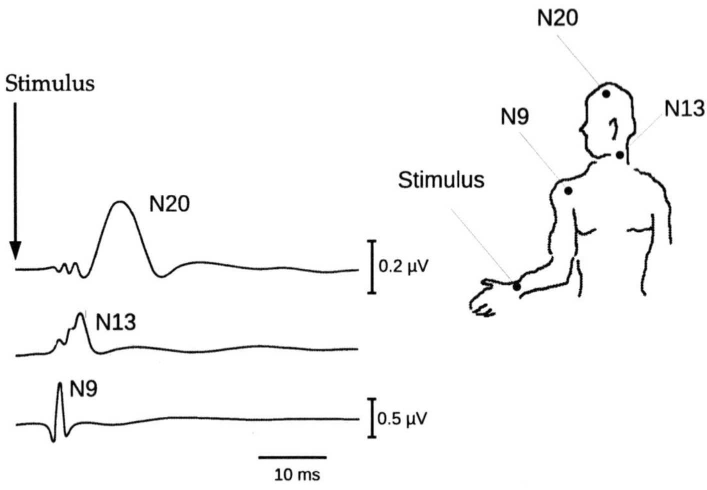

Fig. 6.6 Left: Normal N20 obtained after stimulation of the median nerve at the wrist. Not only is the central response shown, but also the various evoked potentials generated by peripheral neuronal structures. This includes the response generated by the median nerve (N9) and structures in the brain stem (N13).

lingering will show normal SEPs. Finally, in patients suffering from a post-anoxic encephalopathy, absence of the N20 response 72 h after reanimation indicates a prognosis that is invariably infaust[5]. An example of a normal N20 response is shown in Figure 6.6.

6.3.3 Brainstem Auditory Evoked Potentials

Using a click stimulus, it is possible to record several electric field potentials using scalp electrodes. The various potentials are generated by specific brain regions, that form part of the auditory pathways. The auditory pathway starts in the cochlea, where vibrations from sound waves induce small changes in the pressure of the endolymphe, sensed by the various hair cells in the organ of Corti. These hair cells in their turn communicate with bipolar neurons, with cell bodies in the cochlea, projecting to the pontomedullary junction, via the acoustic nerve (N VIII). The entry of the acoustic nerve therefore, is just at the level between the medulla and lower pons. After synapting at this level, second order neurons project to the ipsilateral and contralateral superior olivary nuclei, that are located in the lower pons. Neurons that are located in the superior olive then project to the inferior colliculus and neurons from the inferior colliculus on their turn project to the medial geniculate nucleus of the thalamus, that finally projects to the superior temporal gyrus: the primary

[5] Recent results indicate that a negative SEP after 24 hours also indicates an invariant unfavorable prognosis [7, 8].

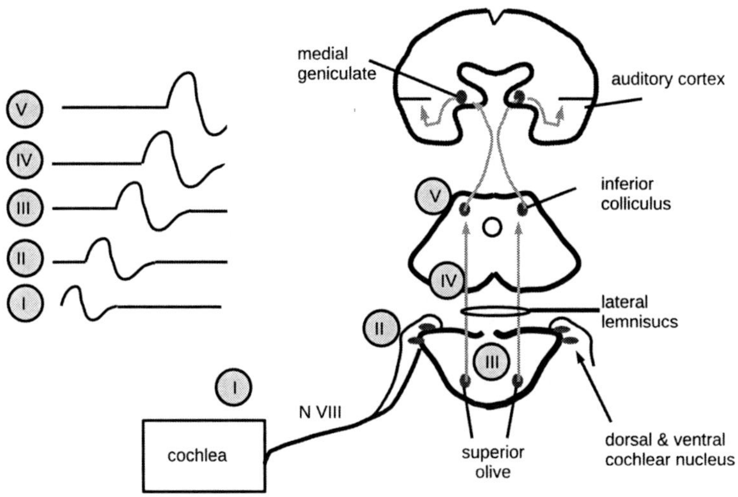

Fig. 6.7 Schematic of waveforms that can be recorded in a BAEP investigation and their sides of generation. The entry of the cranial nerve VIII as at the level of the medulla, just below the pons. The superior olivary nuclei are located more cranially, in the lower pons.

auditory cortex. Note, that acoustic information is transported bilaterally once it enters the central nervous system (from the pontomedullary junction, where the ventral and dorsal cochlear nucleus are located). A schematic of the waveforms and their sides of generation is presented in Figure 6.7. BAEPs are made up of a sequence of five characteristic waveforms or peaks, numbered I-V. Peak I is generated near the eighth cranial nerve, peak II is generated in the cochlear nucleus, peak III in the superior olivary complex and peaks IV and V merge to the lateral lemniscus and inferior colliculus. All these BAEP waveforms represent "far-field" potentials, since they are generated by structures relatively far from the surface recording electrodes. Abnormalities that can be recorded typically relate to increased latencies or absence of waves. Interpretation is fairly straightforward, keeping the anatomy as shown in Figure 6.7 in mind. A short overview is presented in Table 6.1, that also lists the normal latencies. Historically, BAEPs have been studied in humans initially to diagnose suspected lesions of the brainstem or the VIIIth cranial nerve, that could not be visualized by imaging techniques in the pre-MRI area. Investigation of BAEPs have been particularly important in the diagnosis of multiple sclerosis and tumors in the cerebellopontine angle. In addition, BAEPs were most helpful to differentiate between metabolic and structural causes of coma, since in metabolic causes of coma, BAEPs are typically preserved. The introduction of the MRI has significantly decreased the use of BAEPs, since the ease of use and the sensitivity and specificity of this imaging technique is often more than sufficient for diagnostic purposes related

Table 6.1 Patterns of brainstem auditory evoked potential abnormalities and normal (interpeak-) latencies. Reference values obtained from [9].

Finding	mean±SD	Lesion
prolonged I	1.88±0.27 ms	Distal N VIII
prolonged I-III	2.14±0.57 ms	between CN VIII and pons
prolonged III-V	1.93±0.78 ms	lesion between caudal pons and midbrain
prolonged I-V	4.09±0.72 ms	
absent I, normal III, V		mild to moderate peripheral hearing loss
absent III, normal I,V		normal variant
absent V, only		above caudal pons
absence of all waves		severe hearing loss
absence of all waves, except I		brain death

to these pathologies. However, since BAEP measures function, a BAEP can be abnormal before a CT-scan may detect structural changes. In particular, the sensitivity to detect conduction abnormalities for acoustic Schwannoma and other tumors of the cerebellopontine angle varies from 75-100%. Present MRI techniques, however, have similar sensitivities and far better specificities than BAEPs. An important clinical application at present is intraoperative monitoring, e.g. during surgery in the posterior fossa. In addition, BAEPs are used in those patients in whom radiological investigation can be difficult (e.g. neonates, infants, patients suffering from claustrophobia). More details can be found in [10].

6.4 Deep Brain Stimulation

Deep brain stimulation is a technique where electrical impulses are applied to particular brain areas in order to modulate abnormal oscillatory behavior to improve function. The most well-known application is the treatment of Parkinson's disease. However, the technique is increasingly used for other movement disorders, as well, for instance essential tremor and chorea. Other applications include the treatment of seizures, mood disorders (depression) and severe obesity. More details will be discussed in Chapter 10.

Problems

6.1. How much energy is delivered to the tissue in a nerve stimulation study applying a pulse of 1 ms duration, current 40 mA? Use a fair estimate of the skin resistance.

6.2. Proof Equation 6.19.

6.3. Make a simulation in Matlab, where you average subsequent stimulus locked responses. Take random noise as the background activity, with variance σ. Create your own response function, with amplitude R. How many responses do you need to average in order to obtain a signal-to-noise ratio > 10?

6.4. Review the system that conducts the action potential from the primary motor cortex to a peripheral muscle.

6.5. Sketch the anatomical connections from the median nerve to the primary somatosensory cortex. The median nerve also supplies motor fibers, making it a mixed nerve. Can you mention a few muscles that are innervated by the median nerve?

6.6. Sketch the anatomic connections from the retina to the visual cortex.

6.7. Derive Eq. 6.1. Note, that the specific conductivity, $\sigma = l/(R \cdot A)$ with R the resistance. This is straightforward derived from $\rho = RA/l$, with ρ the specific resistivity (also known as specific electrical resistance) in units Ωm. l is a length and A the area.

References

1. Lüders, H.O. (ed.): Deep Brain Stimulation and Epilepsy. Martin Dunitz, an imprint of the Taylor & Francis Group, Abington (2004)
2. Plonsey, R., Barr, R.C.: Bioelectricity: a quantitative approach. Springer, Berlin (2007)
3. Rattay, F.: Modeling the excitation of fibers under surface electrodes. IEEE Trans. Biomed. Eng. 35(3), 199–202 (1988)
4. Roth, B.J., Basser, P.J.: A model of the stimulation of a nerve fiber by electromagnetic induction. IEEE Trans. Biomed. Eng. 37(6), 588–597 (1990)
5. Geddes, L.A., Bourland, J.D.: The strength-duration curve. IEEE trans. biom. Eng. 32(6), 458–459 (1985)
6. Priori, A., Cinnante, C., Pesenti, M., Carp, A., Cappellari, E., Nobile-Orazio, G., Scarlato, G., Barbieri, S.: Distinctive abnormalities of motor axonal-strength duration properties in multifocal motor neuropathy and in motor neuron disease. Brain 125, 2481–2490 (2002)
7. Zandbergen, E.G.J., Koelman, J.H.T.M., de Haan, R.J., Hijdra, A., P.R.O.P.A. C-Study Group.: Sseps and prognosis in postanoxic coma: only short or also long latency responses? Neurology 67(4), 583–586 (2006)
8. Zandbergen, E.G.J., Hijdra, A., Koelman, J.H.T.M., Hart, A.A.M., Vos, P.E., Verbeek, M.M., de Haan, R.J., P.R.O.P.A.C. Study Group.: Prediction of poor outcome within the first 3 days of postanoxic coma. Neurology 66(1), 62–68 (2006)
9. Courjon, J., Mauguiere, F., Revol, M. (eds.): Clinical Application of Evoked Potentials in Neurology. Raven Press, New York (1982)
10. Markand, O.N.: Brainstem auditory evoked potentials. J. Clin. Neurophysiol. 11(3), 319–342 (1994)

Chapter 7
Electromyography

Electromyography literally means "writing of electrical aspects of muscles". In practice, it is aimed at obtaining information about function from measurements from both muscles and nerves; therefore, it is also sometimes referred to as electroneuromyography. Clinical questions related to electro (neuro-) myography include: is the conduction velocity of the nerve within normal limits? Does the muscle contract physiologically if an impulse train enters the terminal axon, releasing acetylcholine to interact at the motor endplate in the neuromuscular junction? Are various reflex-arcs intact? Here, we will treat several general measurement principles, without a discussion of all possible abnormalities that can be found in patients suffering from neuromuscular diseases. These details can be found in standard clinical textbooks, e.g. [1, 2]. However, we will present a few examples to illustrate the techniques.

7.1 Review of Anatomy

We will shortly recapitulate the essentials of the neuroanatomy as relevant in the current context. The motor path from the cortex is schematically illustrated in Figure 7.1. The upper motor neuron makes a synaptic connection to the lower motor neuron, that is located in the anterior horn of the spinal cord. From there, the axon of the lower motor neuron runs to the muscle, joined, in part of his course, by sensory fibers from the posterior root. At the muscle, the motor axon branches and terminates on several muscle fibers. The junction with the muscle fibers is known as the neuromuscular junction. The unit comprised of a single motor neuron, its terminal branches and the muscle fibers it innervates is the "motor unit". The motor unit, therefore, is the elementary unit that realizes muscle contraction. This is illustrated in Figure 7.1, as well. The myelin nerve sheet that surrounds the axons of many motor and sensory nerves realizes the saltatory conduction, significantly speeding op the nerve conduction velocity. This was extensively discussed in Chapter 3. Normal, proper functioning peripheral motor and sensory nerve

M.J.A.M. van Putten: Essentials of Neurophysiology, BIOMED, pp. 133–146.
springerlink.com

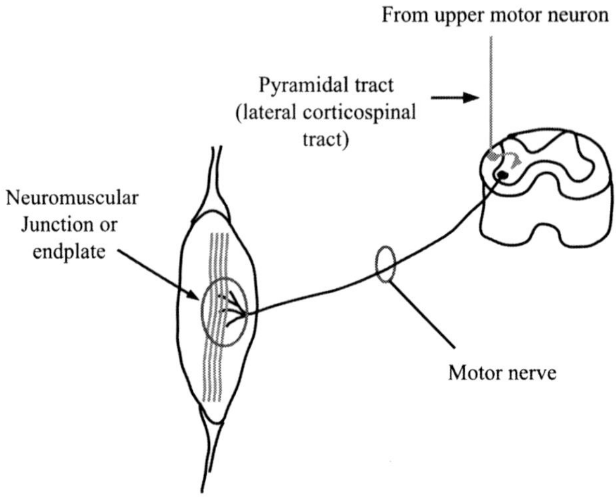

Fig. 7.1 Cartoon of the connections from the upper motor neuron, located in the motor cortex, (only the axon is shown) to the lower alpha motor neuron, located in the spinal cord. The axon from the lower motor neuron connects to several muscle fibers (four in this illustration), via the neuromuscular junction.

fibers show conduction velocities > 40 m/s in the legs and > 50 m/s in the arms.

Various neurological disorders exist that are (partially) reflected by changes in the nerve conduction velocity. A trivial example is a trauma in which a nerve fiber is actually completely cut; clearly, in this case, neurotransmission is impossible, and the conduction velocity is nil. In many other circumstances, however, velocities may decrease, and the nerve function is more or less limited. This group of disorders of the peripheral nervous system is known as neuropathies, discussed previously in Chapter 3, section 3.5. We will shortly discuss neuropathies at a later stage in this Chapter, as well, in relation to the electrodiagnostic examination (section 7.3.).

7.1.1 Neuromuscular Junction

This section discusses the anatomy of the neuromuscular junction, the sequence of events that occurs at the neuromuscular junction in response to a nerve action potential and the role of the acetylcholine receptor channel protein in mediating neuromuscular transmission. The neuromuscular junction is a very special (chemical) synapse, that is abundant in the human body. Here, the neurotransmitter acetylcholine (ACh) is released from the terminal axon in response to an action potential, that subsequently crosses a small distance of 80-100 nm by diffusion to activate a particular ligand-gated ion channel, the acetylcholine receptor. An EM-photograph is shown in

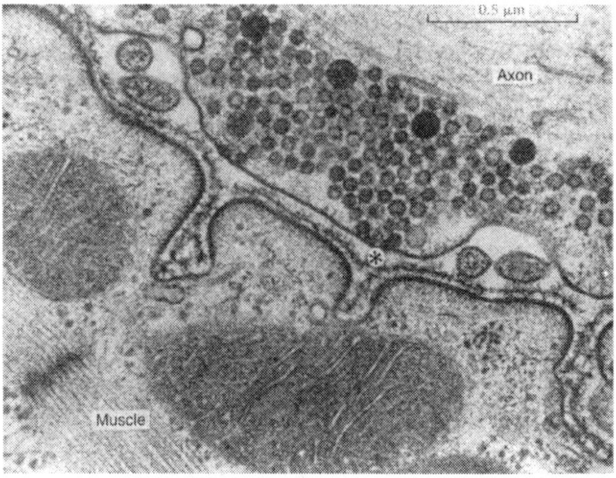

Fig. 7.2 Structural reality of the neuromuscular junction (quick-frozen, freeze-substituted). Courtesy of Dr. John Heuser MD, Heuser Lab, Washington University, St. Louis, USA.

Figure 7.2. Note the large amount of vesicles, loaded with acetylcholine, waiting to be released, at the terminal axon. This anatomical structure of the muscle fiber that belongs to the neuromuscular junction is also known as the motor endplate zone. This is a specialized part of the muscle fiber that contains the various acetylcholine receptors.

In 1972, Katz and Miledi performed classical experiments by making intracellular recordings from a muscle fiber at an endplate at rest [3, 4]. What they found were small fluctuations in the resting membrane potential resulting from the spontaneous opening and closing of ionic channels. In addition, they could observe large deflections, which occurred if a quantum of ACh was spontaneously released from the terminal axon. These large deflections are known as miniature endplate potentials (MEPPs), as illustrated in Figure 7.3.

This release of a quantum of ACh is caused by the spontaneous fusion of a vesicle of ACh, containing 5 to 10-thousand ACh molecules, with the cell membrane. Note, however, that a single vesicle is not sufficient for muscle contraction, since the membrane potential changes only 0.2-0.5 mV. The MEPPs are caused by the nearly simultaneous opening of many ion channels, by activation of various ACh-receptors. Hereafter, ion channels close again. This closing of channels often shows an exponential distribution (Figure 7.3), from which a mean channel open time can be estimated. Recording of MEPPs in a clinical neurophysiology lab is not so difficult. From these recordings, it is now possible to estimate the mean channel open time [5]. In some diseases this mean channel open time is increased, resulting in abnormal functioning

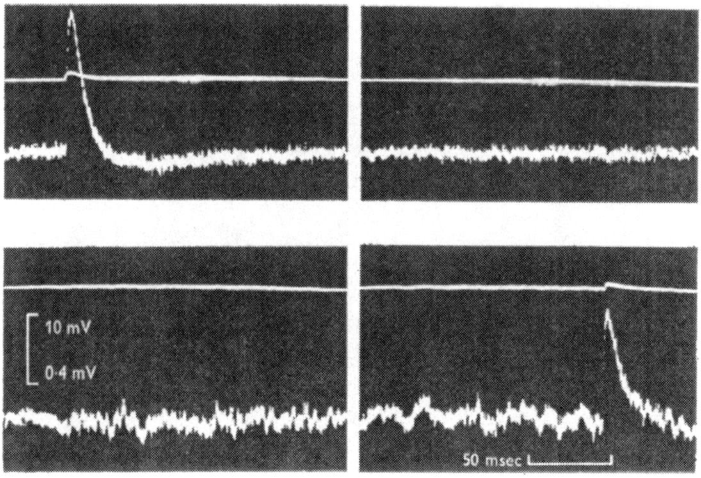

Fig. 7.3 Intracellular recording from an endplate in frog sartorius at 21 degrees C. In each block, the upper trace was recorded on a low-gain d.c. channel (scale 10 mV); the lower was simultaneously recorded on a high-gain condenser coupled channel (scale 0-4 mV). The top row shows controls (no ACh); the bottom row shows membrane noise during ACh application, by diffusion from a micropipette. In the bottom row of records, the increased distance between the low and high gain traces is due to upward displacement of the d.c. trace because of ACh-induced depolarization. Two spontaneous miniature endplate potentials are also seen. Republished from B. Katz and R. Miledi, The statistical nature of the acetylcholine potential and its molecular constituents, J Physiology (1972), 224, pp. 665–699, Blackwell Publishing, Oxford, England, © 1972.

of the neuromuscular junction. An illustration of endplate noise recorded in vivo is shown in Figure 7.4.

If voluntary contraction is performed, various vesicles (quanta) of ACh are released and many ACh-receptors will be activated. This process is limited by diffusion of ACh from the synaptic cleft and the action of acetylcholinesterase, and enzyme that inactivates ACh.

Various neurological diseases exist where there is a failure of the neuromuscular transmission. The most well-known condition is Myasthenia Gravis, an autoimmune disease caused by antibodies that react with the ACh receptor, leaving fewer functional ACh receptors. Another condition is Lambert-Eaton myasthenic syndrome, where the autoimmunity is directed against the presynaptic Ca-channels. These latter are needed to allow excretion of ACh from the pre-synaptic axon terminal, since ACh is released after the opening of the voltage-gated Ca-channels, that allow Ca-influx into the pre-terminal axon. After Ca-influx, fusion of vesicles with the axon membrane becomes possible, and ACh is released. A third disorder is caused by the lack of sufficient

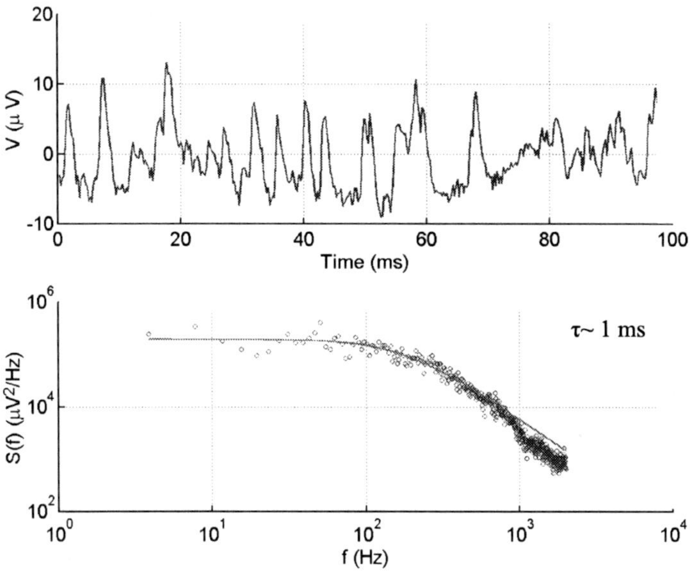

Fig. 7.4 In vivo recording of endplate noise from the Extensor digitorum brevis muscle from the author. In the upper panel, various individual MEPPs are visible. Analysis of this signal, e.g. by fitting a double Lorentzian function to the power spectrum, allows estimation of the mean channel open time (lower panel). Here, the mean channel open time is approximately 1 ms; this is a normal value.

acetylcholinesterase, so called acetylcholinesterase deficiency. We remark that particular nerve gases also work on this structure, for instance by inactivating ACh-esterase, leaving the muscle fibers contracted.

What we have just discussed are some essentials of the neuromuscular junction, in reality there are many other relevant features. And, as holds for almost any structure in the human body, other diseases originating at this structure are known, as well. For instance, some patients suffer from changes in the particular building proteins, such as muscle specific kinase (MUSK). For more details about the neuromuscular junction, we refer to the literature.

7.2 Diagnostic Work-Up

In the diagnostic work-up of a patient suspected of having a particular neuropathy, various characteristics are determined, such as the nerve conduction velocity (m/s) and the amplitude of the responses required from stimulating sensory and motor nerve fibers. Furthermore, the electrical activity from various muscles can be recorded directly, using a small concentric needle electrode. These techniques will be discussed in the next subsections.

7.2.1 *Estimation of the Peripheral Nerve Conduction Velocity*

Our body contains a tremendous amount of nerves: pure sensory nerves, pure motor nerves, mixed nerves and autonomic fibers. Although it is not possible to obtain measurements from all these nerves, many nerves are accessible for examination. Here, we will discuss the general measurement principles for sensory and motor (or mixed) nerves, and illustrate each technique with a clinical example.

Sensory nerve conduction velocity

The conduction velocity of a sensory nerve fiber can be measured by stimulating the nerve at a particular location, and by measuring the time difference between the arrival of the response between the stimulating electrode and recording position. This is exemplified in Figure 7.5.

A clinical example is a patient who may suffer from a carpal tunnel syndrome. In this condition, there is a compression of the median nerve in the carpal tunnel, an area at the basis of the palm of the hand, that is relatively narrow. The (mixed) median nerve needs to pass this tunnel and then gives off several sensory branches to the first, second, third and fourth finger. In

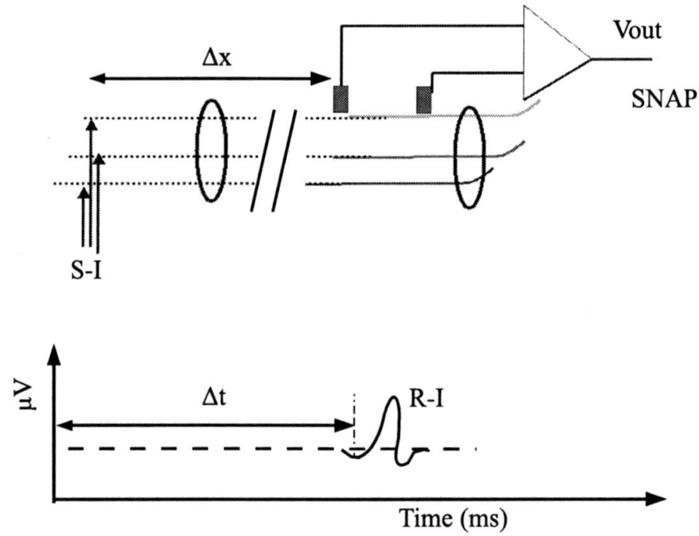

Fig. 7.5 Cartoon of stimulation of and recording from a sensory nerve. Since in this measurement, no synapse is involved, a single stimulation site (S-I) is sufficient for a reliable estimate of the sensory nerve conduction velocity (SNCV) by measuring the time delay, Δt to its response R-I and the distance Δx between the stimulation and recording position.

addition, there is a motor branch to the abductor pollicis brevis muscle. If compressed, patients may complain about tingling sensations in the fingers, and sometimes muscle weakness. Quite typically, many patients wake up during the night from the sensory disturbance. With electromyography, it can often be demonstrated that the conduction velocity from the wrist to the finger is reduced. This measurement is often performed comparing the distal sensory latency (DSL) of the median nerve with the ulnar nerve. This latter nerve gives off a sensory branch to the fifth (and part of the 4-th) finger, but does not run through the carpal tunnel. Both nerves can easily be stimulated at the wrist. By simply comparing the DSL of the median nerve with the DSL of the ulnar nerve, it can be shown that the median nerve DSL is increased in these patients. In severe cases, the SNAP cannot be recorded at all, indicating (almost) complete blockage of sensory nerve transmission.

Motor nerve conduction velocity

Measuring the velocity of a motor nerve is illustrated in Figure 7.6. The actual response that is measured, is called the compound muscle action potential (CMAP). This is a weighted sum of all motor-unit action potentials

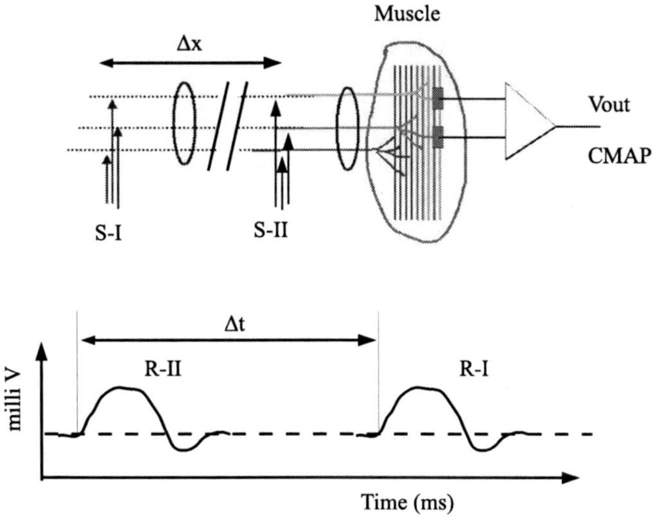

Fig. 7.6 Cartoon of stimulation and recording "from a motor nerve". The actual recording is performed from the muscle innervated by the motor nerve rather than the motor nerve itself. This contrasts the technique employed in estimating velocities from sensory fibers. Note, that in this case, the latency difference Δt between two responses, R-I and R-II, needs to be evaluated, since the synaptic delay at the neuromuscular junction is not exactly known, and may vary between 0.5 to 1 ms. Therefore, stimulation (S-I and S-II) is now performed at two different locations, separated by a distance Δx.

(MUAPs), that are caused by the (near) simultaneous stimulation of the various individual nerve fibers. The amplitude of the CMAP is related to the number of motor units that contribute to the CMAP and their (relative) amplitudes. Therefore, the CMAP amplitude contains information about the number of functioning nerve fibers between the stimulation point and the corresponding muscle. In addition, the number of functioning nerve fibers is relevant.

Because the nerve conduction velocities of the axons in a peripheral nerve are not exactly the same, MUAPs will not be generated at exactly the same moment in the muscle. This asynchronous depolarization of the various muscle fibers due to differences in the conduction velocities of the various axons of the motor nerve is known as temporal dispersion:. Since each MUAP has both a positive and a negative phase, there is partial cancellation of the nett electrical field generated in the muscle. This phenomenon is known as *phase cancellation.* Typically, therefore, the CMAP recorded after stimulation from a more distant site is *lower* than recorded from a more proximal stimulation site.

7.2.2 Needle Examination

Direct electrical examination of muscles is possible by inserting a concentric needle electrode into the muscle. This allows the recording of the electrical activity of one or more muscle fibers, depending on the recording area of the needle. In a normal muscle, there is no "spontaneous" electrical activity *except* if the recording electrode is near the neuromuscular junction, where recording from the endplate zone would be possible, as discussed previously. Therefore, in rest in normal conditions, there are no voltage fluctuations

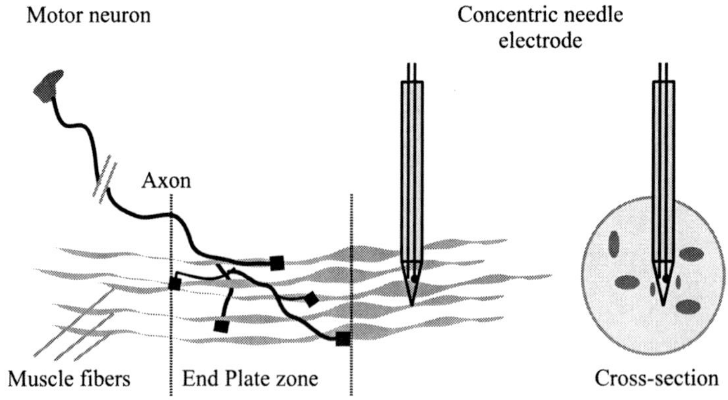

Fig. 7.7 Sketch of the needle electromyography examination. The concentric needle electrode is placed into the muscle, and allows the recording of the electrical activity of various muscle fibers. The illustrated motor unit consists of five motor fibers. At the right a cross-section is shown.

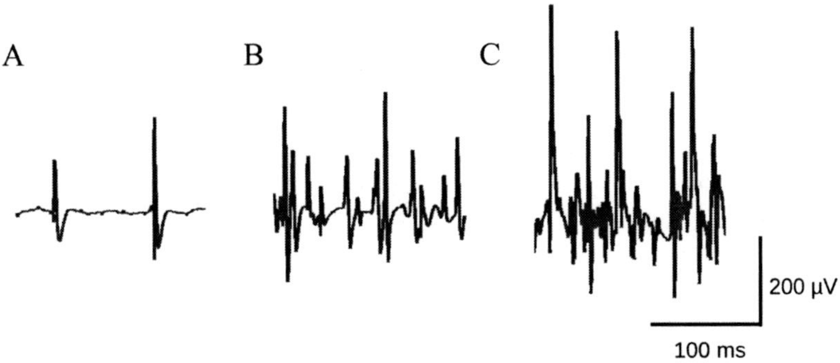

Fig. 7.8 Three different contraction levels. A: two different motor units are shown. Typically, a second unit will fire if the first unit reaches 7-10 Hz. This phenomenon is known as *recruitment*. In the middle illustration (B), much more motor units are visible (mixed pattern). If the muscle contracts even more, a so-called interference pattern emerges: individual motor unit action potentials cannot be seen anymore. This is nearly accomplished in panel C. Data from our own department, as recorded in a healthy volunteer.

recorded (electrical silence) outside the endplate zone. However, if a muscle is (partially) denervated -due to damage of the supplying nerve- *spontaneous pathological muscle fiber activity* will be present. Also, in (primary) muscle diseases (myopathies) abnormal electrical activity is generated[1].

During *physiological contraction* of a muscle fiber, action potentials that travel across the muscle fiber membranes originate at the endplate zone. Here, ACh-receptors are activated by acetylcholine, released from the motor unit terminal branches. If post-synaptic depolarization is sufficiently large, post-synaptic voltage-gated channels will on their be activated and subsequent Ca-influx along the tubuli of the myocyte will eventually lead to muscle fiber contraction. The force of the contraction is basically controlled by two main mechanisms. The first mechanism to increase the muscle force is by increasing the firing rate of individual motor units. The second mechanism is the use of more motor units. In general, force generation starts with relatively small (low force generation) motor units, and with increasing force larger motor units are activated. This process of both increasing the firing rate of individual motor units and the use of an increased number of different motor units is clinically called *recruitment.* If you just barely contract a muscle, a motor unit will fire with approximately 3-5 Hz. If you generate more force, this motor unit will first increase its firing rate up to approximately 10 Hz, and then a second motor unit begins to fire. This is known as the rule of five: the number of active units is approximately the firing rate of the fastest unit divided by 5. This general concepts holds for all voluntary muscles (but in some muscles the ratio's are different. In cranial muscles, single units can fire

[1] These spontaneous discharges encompass fibrillations and positive sharp waves.

up to 15 Hz). During maximal contraction, individual motor units can fire with a frequency of up to 40 Hz. By sufficient training, (visual and auditive) analysis of this recruitment pattern allows differentiation of various abnormal conditions. An example is the patient with a severe lesion of the motor nerve supplying the muscle. If only a few remaining fibers allow nerve-muscle transmission of signals, the recruitment pattern will show only one or a few motor units during maximal contraction, that each fire with their maximum frequency (near 30 to 40 Hz). In a severe myopathy, one can observe "condensed recruitment": very quickly many motor units are activated, but the force generated is much lower than normal.

Another feature that is important in the interpretation of the needle examination is the shape and the number of phases of the motor unit action potential (MUAP). In a healthy muscle, the various fibers that belong to a single motor unit almost all contract at the same moment. In myopathies this synchrony can be disturbed, and motor units show a prolonged discharge, with reduced amplitudes and increased latencies between the initial contraction of the several muscle fibers that belong to this motor unit[2].

7.2.3 Single Fiber EMG

In some neurological conditions, detailed measurements of the neuromuscular junction are relevant. The prime example is myasthenia gravis (MG), where patients suffer from a functional defect at the receptor site of the neuromuscular junction due to antibodies generated against the acetylcholine receptor (AChR).

As you remember, muscle fibers are activated by the opening of sufficient acetylcholine receptors (transporting sodium ions), that on their turn depolarize the muscle membrane. If this depolarization is sufficiently large, there is a propagation of the depolarization along the muscle fiber, activating on its turn voltage sensitive channels, that results in Ca-influx and finally muscle contraction. Typically, this process is very reliable. If action potentials are carried towards the various muscle fibers that are innervated by a terminal axon (that branches off to various fibers: the motor unit), these impulses both arrive at almost the same moment, the thresholds are almost the same and contraction will always occur. However, in patients with myasthenia gravis the thresholds of various muscle fibers may differ, or be so high, that these fibers cannot be stimulated at all.

Suppose it would be possible to measure the time difference between at least two muscle fibers that belong to the same motor unit, we could gain insight into the time interval between the contraction of these muscle fibers. Thanks to pioneering work of Stölberg, needle electrodes have been developed that allow measurement at this anatomical scale. These single-fiber electrodes have a very small "active area", that allows recording from 1-3 single fibers,

---

[2] This results in a more polyphasic MUAP.

only. By stimulating a motor nerve fiber with a particular frequency, say 3 Hz, and subsequent recording of the response of individual muscle fibers to this stimulation, we can record single fiber action potentials (SFAP). The time interval between successive SFAPs is a measure for the variation in the delay between stimulation and response, and is a relevant measure to estimate the reliability of the neuromuscular transmission. In normal muscle, the mean value of these intervals (called "jitter") ranges from 20-30 microseconds. In patients with pre-or postsynaptic transmission disturbances, however, the jitter is increased. Examples include myasthenia gravis and Lambert-Eaton myasthenic syndrome.

7.3 Neuropathies

Neuropathies are all those neurological diseases where there is dysfunction of one or multiple peripheral nerves. Depending on the primary target, i.e. the axon or the myelin that surrounds it, these disorders are known as axonal or demyelinating neuropathies. Neuropathies were discussed previously in Chapter 3. For a definite diagnosis, however, and guiding the treatment of patients, additional tests are typically needed. This includes nerve conduction testing and needle examination. Muscle biopsy and laboratory examination (creatine kinase[3]) may also be useful. The combination of the clinical features, creatine kinase, and the findings of the electrodiagnostic examination can often result in a definite diagnosis. An important differentiation is to distinguish between the (very large group) of axonal neuropathies and the (much smaller group of) demyelinating neuropathies. In an axonal neuropathy, the primary damage is (any part of) the axon. In demyelinating neuropathies, the structure that is initially damaged is the myelin. In clinical reality, however, especially if neuropathies have been present for a longer time, both axonal and demyelinating features will be present. This may imply both a slowing

Table 7.1 Overview of some differences between axonal and demyelinating neuropathies in nerve conduction and needle EMG studies.

Feature	Axonal	demyelinating
MNCV	mild reduction	strong reduction to block
CMAP amplitude	$\downarrow\downarrow$	$\downarrow$
MUAP	polyphasic	can be normal
EMG-rest	fibrillations, positive waves	normal
EMG-contraction	polyphasic MUAP, reduced recruitment	normal MUAP, sometimes reduced recruitment

[3] Creatine kinase is an enzyme that is abundant in muscle tissue. In many myopathies, the serum creatine kinase level is increased.

of the conduction velocities and the presence of phenomena like fibrillations and positive waves. In Table 7.1 we present some typical differences that can be observed in these two different categories with the electrodiagnostic examination.

7.4 Myopathies

A myopathy refers to a diseased muscle, and includes conditions as inflammation (myositis), autoimmune myopathies, and medication induced myopathies (e.g. due to prednisone usage). Patients with myopathies may complain about pain or changes in muscle strength. Electromyography will show normal conduction velocities, but on needle examination various abnormalities can be noted, due to changes in the electrical properties of the diseased muscle fibers. This is both reflected in changes in the form of the MUAPs, as in the recruitment. Interpretation of these biosignals is typically performed by experienced electromyographers.

7.5 Summary of Electromyography

With EMG it is possible to measure various characteristics of nerve and muscle. Action potentials from both sensory nerves (SNAPs), single motor units (MUAPs) and from various motor units (CMAPs) can be determined, and nerve conduction velocities can be measured (see Table 7.2). A very good advanced text on electromyography is [6].

Table 7.2 Overview of some features that can be measured with EMG.

Abbrv.	meaning	unit	normal values
MUAP	motor unit action potential	mV	0.1-2
CMAP	compound muscle action potential	mV	1- 20
SNAP	sensory nerve action potential	µV	5- 40
MNCV	motor nerve conduction velocity	m/s	50 (arm), 40 (leg)
SNCV	sensory nerve conduction velocity	m/s	50 (arm), 40 (leg)

Problems

7.1. A patient complains about muscle weakness in his legs. On neurological examination you find low to absent reflexes; sensation is preserved. On electromyography, the nerve conduction velocity is about 15 m/s, with dispersion.
a) This is normal
b) This is typical for an axonal neuropathy

c) This is typical for a demyelinating neuropathy
d) This is abnormal, but you cannot differentiate between axonal or demyelinating neuropathies.

7.2. A patient complains about loss of muscle strength in her arm. She cannot lift her hand. She tells you that she had a party last night, and fell asleep in a chair. On examination, you find weakness of the wrist extensors. You consider a "Saturday night palsy" due to radial nerve compression. If you perform an electromyography immediately, and you do not find signs of denervation in the muscles innervated by the radial nerve,
a) you can be sure that there is no irreversible damage to the radial nerve
b) you cannot differentiate if axonal damage has occurred, since denervation signs are visible only after 1-2 weeks
c) you will most likely find blocking of nerve conduction at the wrist
e) none of the above

7.3. The following questions are all related to the neuromuscular junction.
a) Sketch a neuromuscular junction
b) Mention two conditions that are characterized by a disturbance of the signal transmission at the neuromuscular junction
c) What ion is conducted by the ACh-receptor
d) Where does botulinum toxin act?

7.4. What is incorrect?
a) A motor unit consists of a terminal axon and all muscle fibers it innervates.
b) Muscle strength is regulated by changes in force of individual motor fibers.
c) Every normal muscle contains muscle fibers from various motor units.
d) Botulinum toxin blocks neuromuscular transmission by interference with acetylcholine release

7.5. What is true?
a) A SNAP is typically higher in amplitude than a CMAP
b) A SNAP is about the same amplitude as CMAP
c) A CMAP amplitude is higher than the SNAP amplitude
d) This depends on the muscle from which the CMAP is recorded.

7.6. You perform a conduction measurement, and estimate a time interval between stimulation and response of 3 ms. You repeat your measurement at a more proximal site, and find a latency of 7 ms. The distance between both stimulation positions is 20 cm. The nerve conduction velocity is about:
a) 50 m/s
b) 55 m/s
c) 40 m/s
d) 42 m/s

7.7. Curare (=dart gun poison from Africa) is a drug that causes paralysis of skeletal muscle. The drug acts at the synapse between the nerve and the muscle cell. Acetylcholine is the neurotransmitter at this synapse. How does curare act at the neuromuscular junction to cause paralysis? Can you suggest some reasons for the use of curare in clinical medicine?

References

1. Levin, K.H., Lüders, H.O.: Comprehensive clinical Neurophysiology. W.B. Saunders Company, Philadelphia (2000)
2. Blum, A.S., Rutkove, S.B. (eds.): The Clinical Neurophysiology Primer. Humana Press, Totowa (2007)
3. Katz, B., Miledi, R.: Membrane noise produced by acetylcholine. Nature, Lond 226, 962–963 (1970)
4. Katz, B., Miledi, R.: Further observations on acetylcholine noise. Nature, New Biol. 232, 124–126 (1971)
5. van Putten, M.J.A.M., Padberg, M., Tavy, D.L.J.: In vivo analysis of end-plate noise of human extensor digitorum brevis muscle after intramuscularly injected botulinum toxin type A. Muscle Nerve. 26(6), 784–790 (2002)
6. Kimura, J.: Electrodiagnosis in diseases of nerve and muscle. Oxford University Press, Oxford (2001)

Chapter 8
Electroencephalography

8.1 Introduction

The electrical activity of the brain was first measured through the intact skull around 1923 by Hans Berger (1873-1941), a German neuropsychiatrist. These recordings were made on his 13-year old son, Klaus, who is now famous because he was the first human being from which EEGs were recorded by a non-invasive method. An example of such a recording is presented in Figure 8.1. It becomes immediately apparent, that the brain generates rhythms.

This recording of the human EEG was a major accomplishment, since modern operational amplifiers, with their high input impedance, where not available at that time. In fact, bioelectrical signals were measured with the string galvanometer (also shown in Figure 5.1). This instrument measured the movement of a conducting wire placed in a static magnetic field as a function of the current through the wire, and the displacement was projected on a moving paper chart. At present, EEGs are recorded using digital equipment, with high-impedance low-noise amplifiers. Digital recording has the additional advantage that subsequent signal analysis is possible in the digital domain, allowing digital filtering and computer analysis. Some of these aspects will be discussed in Chapter 9.

Indications for clinical EEG recordings are varied and include (suspected) epilepsy, metabolic disorders, sleep disorders, brain death, and some forms of neurodegenerative disease, e.g. dementia. EEG monitoring is used in patients during carotid surgery, and in the stroke or intensive care unit. In addition, EEG studies are used world-wide to improve our understanding of various physiological processes in the brain, including attention, memory, autism and language.

M.J.A.M. van Putten: Essentials of Neurophysiology, BIOMED, pp. 147–165.
springerlink.com

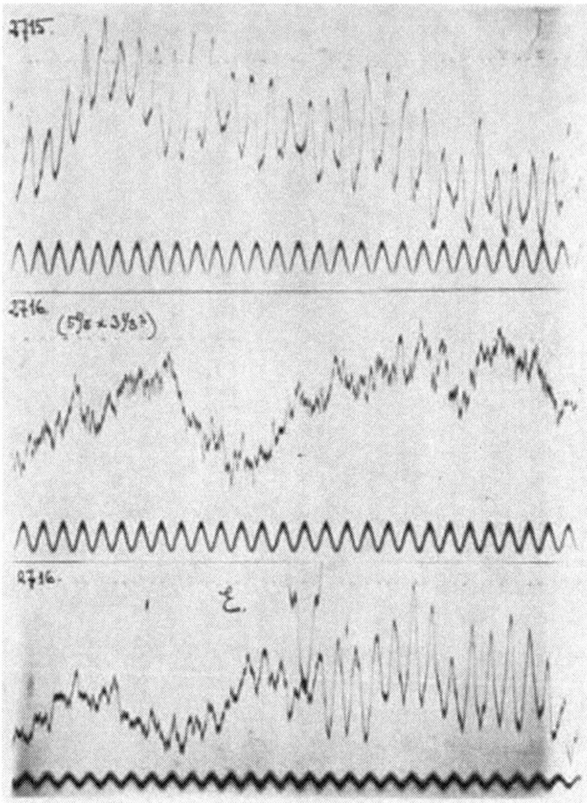

Fig. 8.1 One of the first EEGs as recorded by Hans Berger (1924). The rhythm shown is the α-rhythm. The signal at the bottom of each panel is a calibration signal. Note the attenuation of the α-rhythm by opening of the eyes (middle panel).

8.2 Cellular Basis of EEG Activity

Almost all cells in the human body create voltage differences across their surface; neuronal tissue is specialized for this process. Typically, this voltage is around -70 to -90 mV if the cell is at rest and increases to positive values in the generation of an action potential when a cell becomes excited. These processes have been discussed in detail in former chapters. Let us now look in more detail to a part of the cortex, as shown in Figure 8.2.

This is the basic building structure of the cortex, showing 6 layers. The cortical cells come in various flavors, known as pyramidal cells, various interneurons and glia cells. Some cells have short axons and dendrites (local interactions), some axons are quite long. For instance, cortical motor pyramidal cell have axons that can extend up to the lower part of the spinal cord, which makes the length of the order of a meter. The pyramidal cells in the

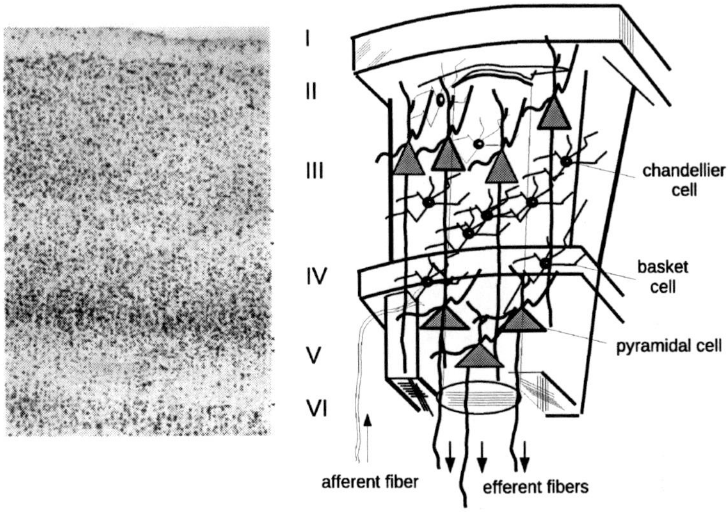

Fig. 8.2 Left: Nissl stain of the cortex, showing the 6 cell layers. At the right, an artist impression of the corresponding cells in the cortex is shown. Note the preferential location of the pyramidal cells in layers II and III (superficial pyramidal cells), and V and VI (deep pyramidal cells).

neocortex consist of two types: superficial pyramidal cells in layers II and III, and the deep pyramidal cells in layers V and VI. In addition to the differences in location, these cells also receive different inputs: the superficial pyramidal cells mainly receive intracortical input; the deep ones receive both intracortical plus subcortical inputs, mainly thalamic[1]. Recall, that the pyramidal cells are aligned at directions perpendicular to the cortex. There are many synaptic inputs to the pyramidal cells, ranging from 5 to 10-thousand per cell. In principle, every event that is associated with changes in the membrane potential of any of these excitable cells (neurons and glia cells) can contribute to voltage changes in the extracellular space. It is primarily the pyramidal cells, however, that contribute to the EEG signal. This is due to their density: pyramidal cells make up 70-80% of all neurons and their dendrites with the synaptic inputs are perpendicular to the skull. To further simplify matters, let us draw a few pyramidal cells and the synaptic input from the thalamus or contralateral cortex. This is illustrated in Figure 8.3. Excitatory input from thalamocortical relay cells ends near the soma of pyramidal cells in layer V and VI. Here, current enters the axon (current sink), while current leaves the axon at the apical part (Figure 8.3-A). Excitatory input from contralateral parts of the cortex has its synaptic connections at these apical parts of the axons (in layer II, Figure 8.3-B), and the current source is now located near the soma. The current dipoles that are thus created have opposite

[1] For more details about the anatomy, see e.g. [1, 2].

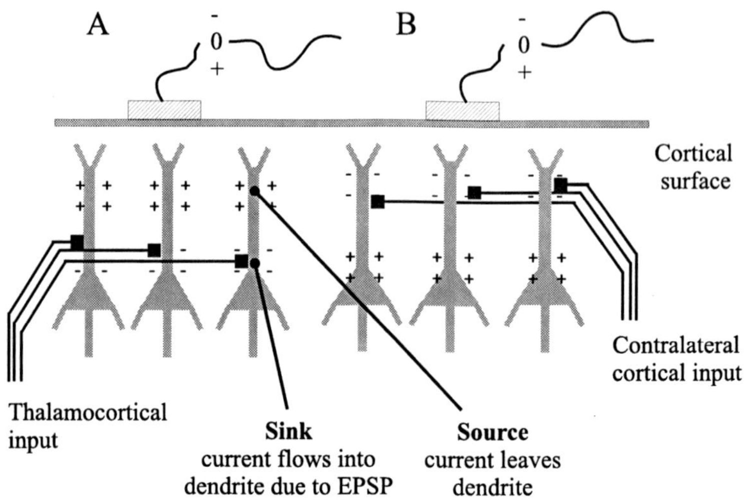

Fig. 8.3 Schematic of six pyramidal cells and thalamocortical, excitatory input (A) and input from the contralateral cortex via the corpus callosum (B). These excitatory synapses depolarize the membrane, and current (direction of positively charges, by convention) enters the axon (current sink). Elsewhere, at the apical parts of the dendrites (in A) or the proximal part of the dendrites, near the soma (B), current leaves the membrane (Kirchoff's law). These synaptic inputs generate current dipoles in a direction perpendicular to the cortex. Redrawn and modified from Kandel et al. Principles of Neural Science, McGraw-Hill Companies Inc, ©2000.

polarity. Thalamocortical, excitatory, input will result in a positive deflection of the voltages measured at the scalp, while excitatory input from contralateral cortex will results in a negative deflection[2].

Of course, the current dipoles of individual pyramidal cells are too small to result in an electrical signal that can reliably be measured at the scalp. However, many cortical pyramidal cells [3, 4] synchronize their activity as a result of synchronous excitatory and inhibitory postsynaptic potentials. This is realized by e.g. coherent input from thalamocortical excitatory input in layer V and VI. The synchronous postsynaptic potentials generated by this afferent input generates a sufficient number of parallel current dipoles. Therefore, neocortical sources of electrical activity can be pictured as *dipole layers.* These dipole layers result in a macroscopic electrical signal at the scalp, the *electroencephalogram.* The voltages that can be recorded at the scalp due to charge variations in this dipole layer, mainly realized by the pyramidal neurons, are of the order of 1-100 µV and consist of frequencies between 0.5 en 35 Hz. Although neuronal sources in the brain do generate higher frequencies than 35 Hz, it is due to the low-pass filtering

[2] For historical reasons, positive deflections are *downward* in neurophysiological measurements and negative deflections are *upward.*

characteristics of the skull and the relative low signal-to-noise ratio of frequencies above 35 Hz that the scalp EEG frequencies are essentially limited to this range[3].

In addition to the synaptic activities of the pyramidal cells, that have the major contribution to the scalp EEG, there are other sources of EEG activity. These are realized both by non-pyramidal cells and non-synaptic events. The nonsynaptic events include fast calcium spikes, voltage dependent oscillations and spike afterpotentials [6]. The action potentials generated by the various neurons are usually not considered to contribute significantly to the scalp EEG. This is mainly due to their short duration (less than 2 ms), the high-pass filtering properties of the extracellular medium, and the (typically) non-synchronous nature of the action potentials. In fact, action potential recordings can only be recorded with micro-electrodes placed very close to the cell body. If action potentials from a larger number of neurons are generated more or less simultaneously (e.g. due to electrical stimulation or in epileptic activity), it is possible to obtain a measurable signal from these "population spikes" with slightly larger electrodes, as well. However, the contribution to the scalp EEG signal is still very small, and in general it can be stated that action potentials are not recorded by the scalp EEG.

8.2.1 Non-synaptic Contributions to the EEG

A nonsynaptic event in neurons that contributes to the EEG is an action potential that is mediated by calcium. These calcium induced spikes are generated in the dendrites, and remain fairly localized, i.e. they do not propagate to the soma. The role of these calcium spikes probably includes the boosting of the synaptic input and the modulation of the synaptic efficiency. Interestingly, part of the "memory function" of the synapse is realized by these calcium currents, as well. The calcium spikes that can be recorded by micro-electrodes represent inward currents in the dendrites, that may occur synchronously with the dendritic EPSPs. Although the quantitative contribution of these dendritic calcium spikes to the EEG potentials is not exactly known, they are probably at least relevant in conditions with increased neuronal synchrony, such as epilepsy.

8.2.2 Contribution of Non-pyramidal Cells to the EEG

Most of the studies on the origin of the scalp recorded EEG, have not examined the role of non-pyramidal cells. The main reason is probably that they are technically more difficult to study. Therefore, their contribution the scalp

[3] This explanation of filtering by the skull is questioned by some, however. See, for instance, Nunez [5].

recorded EEG is not exactly known. The last decade, however, glial cells have received much attention, since they appear to influence the neuronal activity in several ways. For instance, astrocytes in culture respond with increases in intracellular calcium to synaptic stimulation and the application of neurotransmitters. These evoked increases in intracellular calcium propagate as Ca^{2+} waves through gap-junctionally coupled astrocytes, prompting the suggestion that astrocytes can participate in a form of slow, long-distance information processing [7]. Recently, it was shown that astrocytes *in situ* even display *intrinsic* intracellular calcium oscillations that are not driven by neuronal activity [8]. These spontaneous oscillations can propagate as waves to neighboring astrocytes and trigger slowly decaying NMDA receptor mediated inward currents in neurons located along the wave path. These findings show that astrocytes can act as a source for generating neuronal activity in the central nervous system.

In brief, the scalp EEG results mainly from the sum of the various pyramidal current dipoles, both due to their synchronous synaptic activity and their parallel orientation. Therefore, the neocortical sources of electrical activity (at the mesoscale) can be pictured as dipole layers or dipole sheets.

The sensitivity of the EEG to detect changes in the dipole field of the neocortex is a complicated function of the strength of the local dipole fields

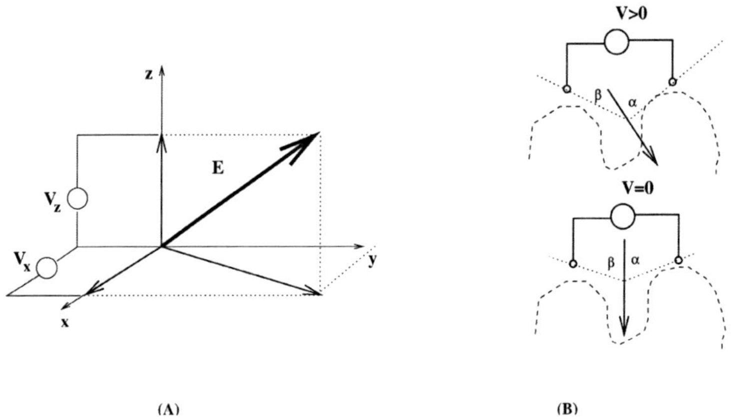

Fig. 8.4 (A) Illustration of the vectorial characteristics of the electric field, E, originating from the neocortical dipole layer. A voltmeter placed at different angles with respect to the vector E will indicate different values. Only by combining several readings, the vector can be reconstructed. In (B) this is illustrated for the measurement of the EEG. The vector shown represents a cortical dipole. In the top illustration, the angles α and β are different and a voltage difference is measured, whereas in the bottom illustration, the two angles are the same: in this case, no voltage difference will be measured. In this latter illustration, we assumed that the dipole is located in the plane of the drawing.

and their orientation. In general, the EEG is most sensitive to correlated (vertical) dipoles in the gyri, and less sensitive to correlated (horizontal) dipoles in the sulcus. In Figure 8.4, the vectorial characteristics of the electrical field are shown and the projection on different axes in a standard 3-dimensional Cartesian space.

In clinical EEG recording it is important to realize that in many (in particular bipolar) measurements, one looks at a particular projection of the neocortical dipole layer. This projection may in some circumstances even obscure important characteristics. For instance, an epileptiform spike (to be discussed later) will not be recorded if the direction of the associated electrical field is exactly perpendicular to the recording direction chosen.

8.3 Recording of the EEG

In the measurement of a biological electrical process, which is based on the flow of ions, it is generally necessary to convert these ionic currents into *electronic* currents, as was extensively discussed in Chapter 5. EEGs are typically recorded using silver-silverchloride electrodes, and a conducting gel, the electrolyte. At this skin-electrolyte-metal interface the conversion of ionic to electronic currents takes place. Most of the power in the scalp EEG is in the frequency range 0.5-35 Hz. For EEG recordings, therefore, it is important that the electrode has a frequency independent characteristic for these low frequencies[4].

As we pointed out earlier, this electrode is also relatively insensitive for movement artifacts. The voltage differences are usually measured between two electrodes (the bipolar measurement), between a single electrode and an "inactive" reference electrode (mono-polar) or between one electrode and the average signal of all other electrodes (common reference).

In a routine EEG recording, the electrodes need be small (to realize a reasonable localization), they must be easy applicable, not cause too much irritation and remain fixated well during the recording period. For short-duration routine recordings (20 min) oftentimes caps are used. If longer recording periods are required, electrodes are "glued" to the skin, that allow long and stable recordings, for over 12 tot 24 hours.

[4] For a typical Ag-AgCl electrode the half cell potential, $E_{\mathrm{hc}} = 0.349$ V, $R_{\mathrm{d}} = 25$ kΩ and $\mathrm{R_s} = 410$ Ω and $\mathrm{C_d} = 4 \times 10^{-8}$. For symbols, see Figure 5.5. It is easy to verify, that the impedance is nearly constant up to frequencies of 35 Hz, using

$$Z = R_s + \frac{R_d}{1 + \omega^2 R_d^2 C_d^2}. \tag{8.1}$$

The student is advised to evaluate that this indeed results in a nearly constant impedance for frequencies below 30-35 Hz.

8.3.1 Recording Sites

In clinical practice, the EEG is usually recorded using the international 10-20 system for standardized electrode placement. The nomenclature of each electrode site follows an alphabetical abbreviation that identifies the lobe or area of the brain to which it refers: F stands for frontal, Fp for frontopolar, T for temporal, C for central, P for parietal and O for occipital. The subscripts that follow refer to the specific location of the electrode within the hemisphere. Even numbers refer to the right side of the head, odd numbers to the left side. The subscript"z" (zero) points to electrode sites in the midline of the skull. As an example, T4 is located over the right temporal lobe and F3 over the left frontal lobe. Viewed from above, the electrodes are now positioned as indicated in Figure 8.5. For the reference signal in the actual recording, it is most convenient to use the mean value of all electrodes (G19). The advantage of recording each signal referenced to this G19 reference before digitization, is that afterwards (in the digital domain) many different *montages* can be realized. For instance, to realize the voltage difference between F7 and T3, one takes F7-G19-(T3-G19) = F7-T3. A montage is a particular combination of channels to display a number of all the possible voltage differences, typically between 19 and 21. In the longitudinal bipolar montage, voltage differences are measured between neighboring electrodes in the anterior to posterior direction, as illustrated in Figure 8.5. This montage has a fair emphasis on symmetry between homologous regions and mainly measures voltage differences in the longitudinal direction.

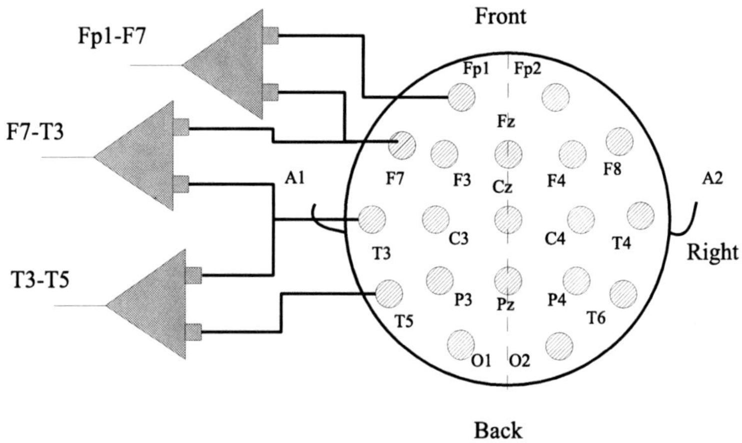

Fig. 8.5 The 10-20 system for the placement of the electrodes in the recording of the standard EEG. A part of a bipolar montage is shown on the left.

In the source montage, or Laplacian derivation, the voltage at a particular electrode is referenced to the mean value of the four nearest neighbors[5]. For instance, the voltage displayed at electrode Cz is then given by

$$U_{\mathrm{Cz}} = Cz - G19 \cdot \frac{1}{4}\left[C3 + C4 + Fz + Pz - 4 \cdot G19\right]. \quad (8.2)$$

In clinical reviewing of the EEG, the electroencephalographer typically switches between different montages for the final interpretation. This is especially necessary for defining the direction of the electric field.

8.4 Main Characteristics of the Human EEG

The main characteristics that are identified in a normal EEG will be shortly discussed. The first thing that is noted, is the presence of rhythms[6]: more or less regular oscillations in the potentials recorded, in the frequency range 0.5 to 35 Hz. The second characteristic of these rhythms is, that they are fairly symmetric between the two hemispheres, with amplitude gradients in the anterior to posterior direction. The frequency range and location of several EEG rhythms are presented in Table 8.1.

Table 8.1 EEG rhythms, their dominant frequency and primary location during wakefulness in adults.

rhythm	frequency (Hz)	primary location	suppression by
α	8-13[a]	PTO	eye opening, mental activity
μ	8-12[b]	central/motor cortex	contralateral movement
β	12-25	frontocentral[c]	-
θ	4-8	(fronto)temporal	-
δ	0.5-4	should be absent (awake)	
γ	30-40	see footnote (d)	

[a] Occasionally, the α-rhythm has a frequency exactly the half or double of the typical α-rhythm in the frequency range 8-13 Hz; these are so called α variants, with a similar reactivity to eye opening.
[b] The μ-rhythm is the rhythm with a comparable frequency as the α-rhythm (though, often slightly higher), but is i) located mainly over the central regions and ii) shows a different waveform, resembling the Greek letter μ.
[c] If present diffuse, this is most often due to the use of benzodiazepines.
[d] γ oscillations play a role e.g. in conscious perception, but are difficult to record or observe in routine EEG.

[5] This is one of the possible implementations. It is also possible to take all electrodes into account, and to weigh the signals with the distance from the "active" electrode.

[6] A fascinating book about the relevance of brain rhythms is "Rhythms of the Brain" by György Buzsáki [9].

Reactivity refers to the change of rhythms by a particular stimulus. For instance, the α-rhythm is modulated by the opening or closing of the eyes. Also, mental action may modulate the α-rhythm. Other rhythms show changes by particular stimuli or actions, as well. For instance, the μ-rhythm, that can be recorded over the motor cortex, is modulated by performing particular movements. In addition, this rhythm may also modulate by merely thinking about a movement or looking at a particular movement[7]. In comatose patients, EEG changes before and after or during auditory or pain stimuli can be evaluated, indicating (some preservation of) central processing.

The suppression of the α-rhythm by eye opening and the suppression of the μ-rhythm by motor activity is caused by "desynchronization" in neuronal populations. If neuronal assemblies are"idling", i.e. waiting to perform an action, they often generate more or less regular rhythmic activity[8]. However, when engaged in some form of processing, interactions in these neuronal populations changes, eventually resulting in differences in scalp current densities and voltages, reflected in modulation of EEG rhythms.

In neonates and children, EEG rhythms are very different from the adult EEG. For instance, the adult alpha frequencies appear around the age of 2 years. Before this age, the EEG is rather polyfrequent, and the dominant rhythm over the occipital areas is slower, with frequencies between 3-8 Hz, depending on age.

Besides the various rhythms, we can identify different *wave forms*, that constitute these rhythms. For instance, the α-rhythm is often sinusoidally shaped, whereas the μ-rhythm is somewhat shaped like the Greek μ. Besides these rhythms and wave forms, it is important where and how often the several rhythms and their particular wave forms are present.

In Figure 8.6, two fragments of a normal EEG, recorded in an adult, are shown, one with eyes closed (EC) and one with eyes open (EO). Note the attenuation of the α-rhythm between the two recordings. The frequency of the α-rhythm is approximately 8.5 Hz in this healthy volunteer. There is a good symmetry between the rhythms from the left and right hemisphere.

All these characteristics of the EEG belong to the "background pattern", and define its mean statistical characteristics. In addition, transients can be present, as can for instance be observed in patients suffering from epilepsy. We have only discussed a few aspects of the normal human EEG. For interested students, good textbooks that provide a much more in depth treatment are [6, 11, 12].

---

[7] Here, mirror neurons are likely involved. Mirror neurons were discovered in monkeys by Giacomo Rizzolatti. In man, these neurons are most likely present, as well, and presumed relevant in e.g. (motor) learning and empathy [10].

[8] In some pathological conditions, there is an increase in the α-activity in the frontal regions. This indicates that the frontal areas are "idling", - which may e.g. occur in Alzheimer's disease or during mild sedation. In normal conditions, this is hardly ever observed in an awake individual, with eyes open.

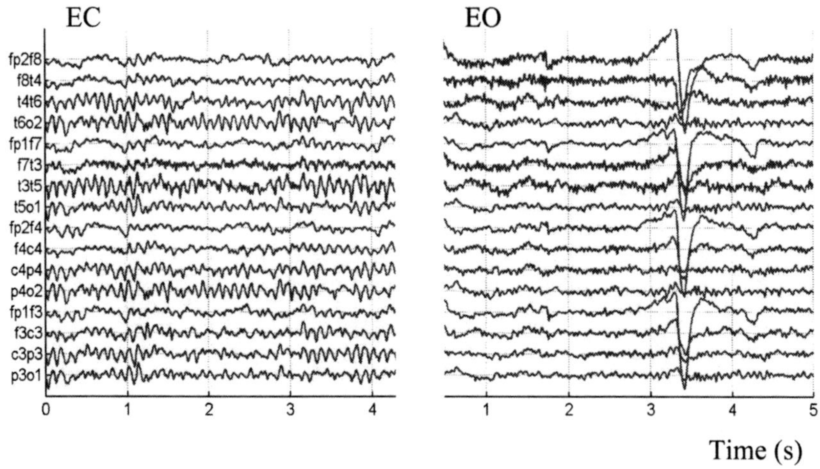

Fig. 8.6 Fragments of a normal EEG in a bipolar montage. Initially (left), the subject has his eyes closed (EC). The right epoch shows the EEG with eyes open (EO). Note the attenuation of the α-rhythm. The phenomenon occurring at around t=3 s at the right is an eyeblink artifact.

8.5 EEG in Epilepsy

For patients with a suspected seizure disorder (epilepsy), the routine EEG is used by clinicians to identify the presence of interictal epileptiform discharges. These discharges serve as a marker for epilepsy, and signify the increased tendency of the brain to generate seizures. Normal EEG findings, however, do not exclude the possibility of epilepsy, due to its limited sensitivity of about 30-50% in the interictal period[9].

Epileptiform discharges are paroxysmal EEG waveforms that have a high degree of association with seizures. In fact, these epileptiform discharges are a signatures of an abnormal (typically increased) synchronization of neuronal populations. The common types are (poly)spikes, sharp waves and spike-and-wave discharges, but other waveforms can be observed as well. In fact, there are quite some signatures of epilepsy in the EEG, which makes the interpretation not always easy. A few examples of these waveforms are presented in Figure 8.7. Spikes are potentials with a short duration (50-70 ms) and sharp waves have a duration of 70-200 ms. A spike-wave-discharge consists of a spike followed by a slow wave. These interictal EEG discharges seen in routine EEG tracings usually occur singly or sporadically. This is one of the reasons why the sensitivity of a routine EEG recording for epilepsy is limited:

[9] This holds for our current "standard" analysis of the EEG, i.e. mainly visual. Some studies using quantitative EEG features suggest that differences in the interictal EEG between normal individuals and patients with epilepsy may exist.

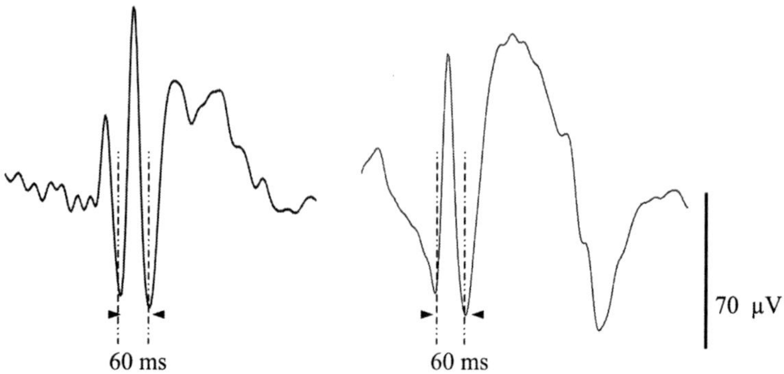

Fig. 8.7 Illustration of two spike-wave complexes, typical epileptiform EEG discharges. The duration of the spikes is 50-70 ms, here about 60 ms. The left waveform actually consists of two spikes, before the slower wave appears.

in patients with epilepsy, the brain often generates normal rhythms, and only now and then interictal epileptiform discharges. The time interval between these discharges can vary from minutes to hours or even to days. Therefore, during a single routine 20 min EEG recording these interictal events are often not present.

In contrast, ictal EEG changes usually consist of repetitive or periodic activity that may show a rather abrupt onset that are associated with a particular clinical manifestations of a seizure[10]. Depending on the spread of these ictal discharges, focal or generalized epileptiform activity is present. An example of a generalized seizure is presented in Figure 8.8. Natural questions to ask are: what are the underlying mechanism responsible for the various epileptiform waveforms, either interictal or ictal and how is it possible that a healthy brain can become epileptogenic, for instance after a trauma or brain infarction? This process of epileptogenesis is quite complicated indeed, and many factors may contribute to abnormal firing patterns and synchronization of neurons, resulting in seizures. Contributions include alterations in the ratio of excitatory and inhibitory neurotransmitters, channelopathies and selective cell damage. Genetic factors may also be involved. A well-known example are the benign familial neonatal convulsions (BFNC), that are caused by an abnormal potassium channel. These children initially suffer from seizures, but when they grow older, the potassium channel is replaced by the adult form, which functions properly; at the same time, the epilepsy disappears. In addition to clinical research and studies that are aimed at improving our understanding of the function of various ion channels in epilepsy, there is an increase of computational studies [14]. Due to our improved insight into cell morphology, location of various ion channels and the kinetics of individual

[10] In long lasting seizures, multiple patterns of recurrent brain states may occur [13].

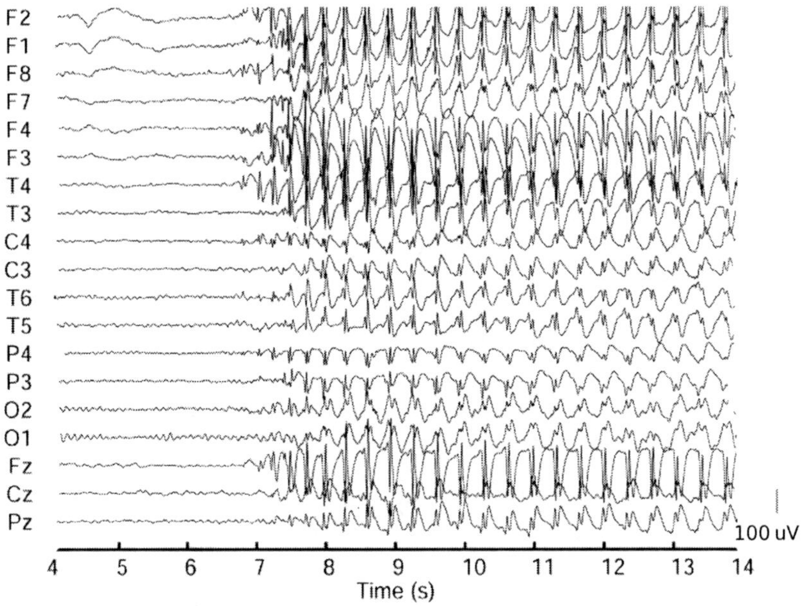

Fig. 8.8 Illustration of a generalized absence seizure, starting at ~ 7 s. There is an abrupt transition to generalized high-amplitude events, characterized by rhythmic (approximately 3 Hz) spike-wave discharges. This is an EEG from a young patient (6 years old) with absence epilepsy, a relatively mild form of epilepsy, where the clinical manifestations are characterized by relatively short periods in which consciousness is lost. In most cases, muscle tone remains present. Clinically, these patients sometimes appear to be day dreaming, while they are in fact having a seizure at that moment.

ion channels, and the continuous raise in computational power, we are now able to create computer models of neuronal function with increasing reality[11]. These computer simulations have proposed various mechanisms that are relevant in epileptogenesis. As a short example, in the genesis of the generalized spike-and wave patterns in absence epilepsy (i.e. the phenomenon shown in Figure 8.8) the thalamocortical network is essential for its generation. If the intracortical fast inhibition (mediated by GABA receptors) is reduced, spike-and wave patterns can be generated. During the "spike", cells fire prolonged high-frequency discharges, followed by slow waves that are coincident with slow potassium currents [15]. These models are also potentially valuable for the diagnostic process and to guide treatment of various diseases. For instance, adding a particular drug can be simulated by changing the strength

[11] Computational neuroscience has found very wide application, by now. In addition to epilepsy, topics include sleep, memory, stroke, etcetera. One of the most large-scale projects is the Blue Brain Project in Lausanne, `http://bluebrain.epfl.ch/`. This project is a joint effort of EPFL and IBM to reverse engineer the mammalian brain.

of excitatory or inhibitory connections; this allows us to study its effect on neuronal rhythms and function. In addition, these models may assist in the design of clinical trials. It is beyond the scope of this book to discuss these interesting matters in more detail. Further references are e.g. [12, 16, 17].

8.6 EEG in Ischaemia

Neurons need oxygen and glucose for proper function, and deprivation will quickly result in a loss of activity. Due to the intimate relationship that exists between neuronal function and neuronal rhythms, oxygen and or glucose deprivation is directly reflected in changes in EEG rhythms [18]. This is typically manifested as slowing of rhythms, with a decrease of fast β-activity and an increase in delta waves. The first neuronal process that fails in moderate ischaemia is the synaptic transmission. The brain uses approximately 30% of its metabolic demand for the excretion of neurotransmitters (and recycling). Since the EEG directly reflects the EPSPs and IPSPs of pyramidal neurons, it comes as no surprise that processes that interfere with synaptic excretion immediately influence the various EEG rhythms. This characteristic is for instance used in monitoring the brain in carotid endarterectomy. This procedure is performed in patients suffering from a significant symptomatic stenosis in one of their carotid arteries. A possible clinical manifestation of

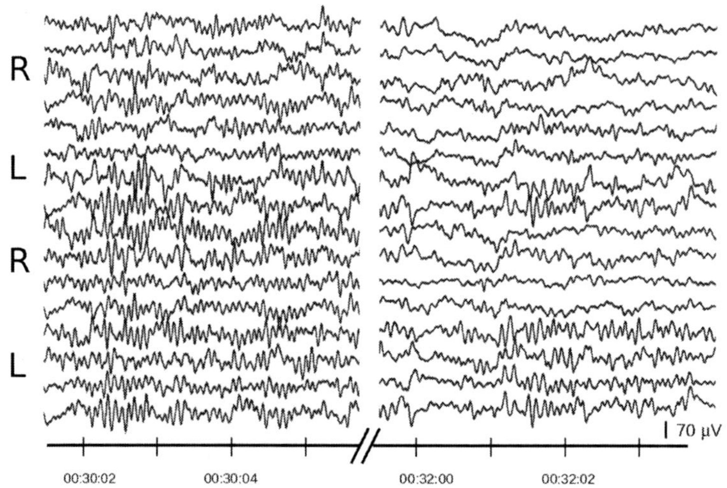

Fig. 8.9 EEG recording during carotid endarterectomy. At the left, the EEG is symmetric. During test-clamping (right) a significant asymmetry is noted (slowing and amplitude reduction of rhythms from the right hemisphere) that warrants the placement of a temporary shunt. The montage shown is a bipolar montage. Traces 1-4 and 9-12 are recorded from the right hemisphere (R), the remainder are recorded from the left hemisphere (L).

such a stenosis is a transient ischaemic attack in the vascular territory of this vessel, for instance a transient dysphasia if it concerns the perfusion of the dominant hemisphere. To reduce the risk of recurrence, the vascular surgeon can perform an endarterectomy, removing the stenosis. This is often performed under general anesthesia, and during the actual removal of the atheroslerotic plaque, the artery needs to be temporarily closed (clamped). To verify that there is still sufficient perfusion of the cerebral hemisphere, an EEG is recorded continuously, and during the test-clamping procedure, one evaluates if asymmetries (or diffuse slowing) occurs. If this is the case, there is apparently insufficient blood supply from collateral vessels, and it is necessary to place a temporary shunt to ensure sufficient blood flow to the brain[12]. An example of unilateral slowing during this test-clamping procedure is presented in Figure 8.9.

8.7 EEG in Sleep

Sleep has a profound effect on the EEG rhythms, and EEG recordings are widely used to assess the quality of sleep in patients suffering from various sleep disorders, including insomnia, sleep apnea syndrome and narcolepsy. In the evaluation of potential sleep disorders, it is not only the EEG that is being recorded. Typically, oxygen saturation, respiratory flow, sound, abdominal and thoracic excursions, and EMG (from the chin and legs) are measured as well, to be well-informed about desaturations, apnea (cessation of respiratory flow for more than 10 s) or periodic leg movements.

Sleep is clinically divided into 3 stages of non-REM (NREM) sleep: stage I, somnolence or "drowsy sleep", stage II and deep sleep (stage III). In addition to these three non-REM sleep stages, sleep is characterized by the existence of REM sleep. In the transition from wakefulness to NREM sleep, initially the spatial gradients of the EEG rhythms become less, and the rhythms slow down to theta waves (stage I). In sleep stage II, beta spindles and K-complexes are present, while in deep sleep (stage III) the EEG shows mainly slow wave activity in the delta range. REM-sleep (dreaming) is characterized by the presence of rapid eye movements (REM) and an EEG pattern that superficially resembles the awake EEG. During sleep, cycles are observed, typically from stage I to stage II and III, back to II followed by a period of REM sleep. This cycle is, on the average, 90-130 minutes in duration. In the beginning of the night, there is more stage III sleep, later in the night there is more REM sleep. An example of a sleep pattern (known as a hypnogram) is shown in Figure 8.10.

[12] One can wonder, why shunting is not performed at all times. The rationale for selective shunting is that placing a shunt bears some risks, as well. Therefore, selective shunting is the common procedure.

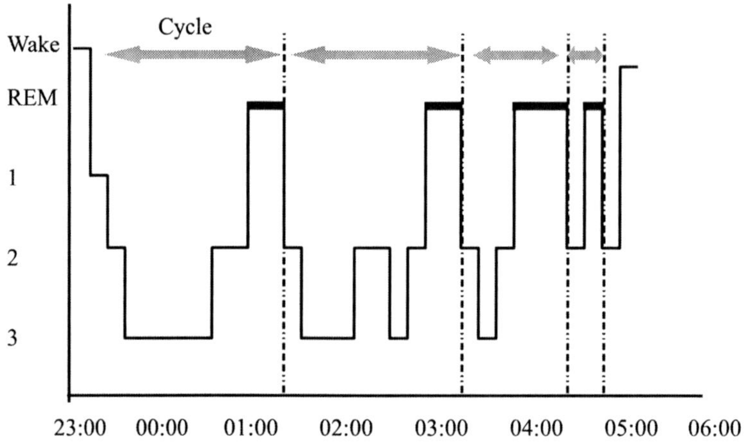

Fig. 8.10 Hypnogram, illustrating the cyclic pattern in sleep. Four cycles are shown. The time on the x-axis shows hours.

8.8 EEG in Brain Death

In assessing brain death, the EEG is an important diagnostic tool. In particular in those patients that are in the Intensive Care Unit, and could be candidates for donation, it is necessary to determine if brain death is present or not. Although other means can be employed as well (e.g. ultrasound and angiography to evaluate if there is cerebral perfusion left), the EEG is still the standard technique to determine if there is any cerebral activity. For the interpretation, it is important to know if the patient is given any sedative drugs and if the temperature is above 32 degrees Celsius. If this is not the case, reliable interpretation is not possible. However, if a reliable recording can be made, and, at a sensitivity of 20 µV per cm, no signs of electrical cerebral activity are observed for more than 30 minutes, a patient is declared brain dead[13]. For more details, see e.g. [11].

8.9 Quantitative Analysis

At present, the standard procedure for the interpretation of the EEG is by visual analysis. Waveforms are evaluated by a clinical neurophysiologist, well-trained in the interpretation of the various EEG rhythms Given the large number of potential manifestations of abnormal EEG patterns, a relatively long learning curve is needed in order to be able to interpret the clinical EEG. Drawbacks of this approach (besides the long learning curve)

[13] In many countries, including the Netherlands, a patient is declared brain dead if the brain stem function if absent, as well. This can be evaluated with various reflexes (e.g. corneal, oculocephalic) and the absence of a respiratory drive.

are the inherent subjective elements that are involved in such a process. Furthermore, it is well-known that human interpretation may suffer from both intra- and interobserver inconsistencies. In addition, the description of the various waveforms is rather phenomenological, using language to communicate various features of the time-series. This has its limitations, as well, since language is not always suited to describe particular elements from the EEG. This has motivated various researchers to explore if the computer can assist to extract relevant EEG features and/or to transform the EEG to alternative domains [19, 20, 21, 22]. This includes the detection of asymmetries, epileptiform discharges or the analysis of sleep patterns. We will present a few examples of quantitative EEG techniques in the next chapter.

8.10 Summary

This chapter is a brief introduction about the generation and measurement of the electroencephalogram (EEG). It discusses some general characteristics of the normal EEG. Various applications of the EEG in clinical neurology are presented, including the role of the EEG in epilepsy, neuromonitoring and sleep. Students who wish to learn more about the basics of the EEG, should read *Electrical Fields of the Brain- The neurophysics of EEG*, by Paul Nunez and Ramesh Srinivisan [5]. This is an excellent text -to become a classic in the field of EEG- that presents both clinical and theoretical information regarding the electrical phenomena of the brain.

Problems

8.1. The EEG measured at the scalp is mainly generated by
a. action potential trains
b. currents induced by EPSPs and IPSPs in interneurons
c. currents induced by EPSPs and IPSPs in pyramidal cells
d. all of the above

8.2. What is a montage?

8.3. What are typical voltages measured with scalp EEG recordings?

8.4. Sketch a typical power spectrum of the EEG, as measured over the occipital areas during an eyes closed condition.

8.5. What is meant by attenuation of the α-rhythm?

8.6. Between which frequencies can you find delta waves? In what conditions can delta waves be present in healthy persons?

8.7. Mention some diseases that may induce left-right asymmetries in the EEG.

8.8. You are asked to develop a quantitative technique to measure left-right asymmetries of the EEG, that needs to be displayed as a single number. Discuss various possibilities. You are allowed to use any source of information.

8.9. What is an interictal EEG?

8.10. Which EEG abnormalities can be found in patients suffering from epilepsy, and what do they represent?

References

1. Kandel, E.R., Schwartz, J.H., Jessell, T.M. (eds.): Principles of Neural Science. McGraw-Hill, New York (2000)
2. Fitzgerald, M.J.T., Gruener, G., Mtui, E.: Clnical Neuroanatomy and Neuroscience. Elsevier Saunders (2007)
3. Creutzfeld, O., Watanabe, S., Lux, H.D.: Relations between EEG phenomena and potentials of single cortical cells.i. evoked responses after thalamic and epicortical stimulation. Electroencephalogr. Clin. Neurophysiol. 20, 1–18 (1966)
4. Petsche, H., Pockberger, H., Rappelsberger, P.: On the search for the sources of the electroencephalogram. Neuroscience 11, 1–27 (1990)
5. Nunez, P.N., Srinivasan, R.: Electric Fields of the Brain. The Neurophysics of EEG. Oxford University Press, Oxford (2006)
6. Ebersole, J.S., Pedley, T.A. (eds.): Current Practice of Clinical Electroencephalography. Lippincott Williams & Wilkins (2003)
7. Dani, J.W., Chernavsky, A., Smith, S.J.: Neuronal activity triggers calcium waves in hippocampal astrocytic networks. Neuron 8, 429–440 (1992)
8. Parri, H.R., Gould, T.M., Crunelli, V.: Spontaneous astrocytic Ca^{2+} oscillations in situ drive NMDAR-mediated neuronal excitation. Nature Neuroscience 4(8), 803–812 (2001)
9. Buzsáki, G.: Rhythms of the Brain. Oxford University Press, Oxford (2006)
10. Rizzolatti, G., Craighero, L.: The mirror-neuron system. Annu. Rev. Neurosci. 27, 169–192 (2004)
11. Levin, K.H., Lüders, H.O.: Comprehensive clinical Neurophysiology. W.B. Saunders Company (2000)
12. Niedermeyer, E., Lopes da Silva, F.: Electroencephalography: Basic principles, clinical applications and related fields, 4th edn. Lippincott Williams & Wilkins (1999)
13. van Putten, M.J.A.M., van Putten, M.H.P.M.: Discovery of recurrent multiple brain states in non-convulsive status epilepticus. Clin. Neurophysiol. 118(12), 2798–2804 (2007)
14. Soltesz, I., Staley, K. (eds.): Computational Neuroscience in Epilepsy. Academic Press, London (2008)
15. Destexhe, A.: Spike-and-wave oscillations based on the properties of GABAB receptors. J. Neurosci. 18(21), 9099–9111 (1998)
16. Schwartzkroin, P.A. (ed.): Epilepsy: models, mechanisms and concepts. Cambridge University Press, Cambridge (1993)
17. Pitkänen, A., Schwartzkroin, P.A., Moshé, S.L. (eds.): Models of Seizures and Epilepsy. Elsevier Academic Press, Amsterdam (2006)

18. van Putten, M.J.A.M., Peters, J.M., Mulder, S.M., de Haas, J.A.M., Bruijninckx, C.M.A., Tavy, D.L.J.: A brain symmetry index (BSI) for online EEG monitoring in carotid endarterectomy. Clin. Neurophysiol. 115(5), 1189–1194 (2004)
19. Gotman, J., Flanagan, D., Rosenblatt, B., Bye, A., Mizrahi, E.M.: Evaluation of an automatic seizure detection method for the newborn EEG. Electroencephalogr. Clin. Neurophysiol. 103(3), 363–369 (1997)
20. van Putten, M.J.A.M.: Nearest neighbor phase synchronization as a measure to detect seizure activity from scalp EEG recordings. J. Clin. Neurophysiol. 20(5), 320–325 (2003)
21. van Putten, M.J.A.M., Tavy, D.L.J.: Continuous quantitative EEG monitoring in hemispheric stroke patients using the brain symmetry index. Stroke 35(11), 2489–2492 (2004)
22. van Putten, M.J.A.M.: The colorful brain: visualization of EEG background patterns. J. Clin. Neurophysiol. 25(2), 63–68 (2008)

Chapter 9
Digital Signal Analysis

9.1 Introduction

The goal of (digital) signal analysis is to extract relevant information from the signal, in order to increase our knowledge of the system that generates the signal. Examples include the detection of a dominant frequency in the EEG or a particular transient (e.g. an epileptiform discharge). Signal analysis is often a multi-step process. For instance, initially filtering may be needed, artifact rejection or averaging. These are all procedures to increase the signal-to-noise ratio, which may eventually result in a signal that allows reliable interpretation. In many circumstances, the biosignal that is acquired fluctuates as a function of time. These kind of signals are known as time-series, and the techniques applied are sometimes referred to as (digital) time-series analysis.

Particular characteristics of these time-series can be described by various features, for instance by the mean and the variance. This is the field of descriptive statistics. It is also possible, if there is an adequate model that describes the signal, to identify the model parameters: parameter estimation. The choice of a particular approach is a function of many "variables". This includes not only the signal characteristics itself or the presumed essentials of the generator of the signal, but also the familiarity with a particular technique or the amount of computing power available. The number of ways with which particular signal features can be extracted is overwhelming [1, 2, 3], and it should not be forgotten that classical visual analysis of biosignals has been shown to be very useful, as well [4].

This chapter will provide a concise review of various basics, which are relevant for digital signal analysis, with a strong emphasis on the analysis of the EEG. We discuss analog to digital conversion and practical issues related to the spectral analysis of neurophysiological signals. For a more in-depth treatment, the reader is referred to other textbooks that are dedicated to digital signal processing, e.g. [5].

M.J.A.M. van Putten: Essentials of Neurophysiology, BIOMED, pp. 167–184.
springerlink.com

9.2 Analog to Digital Conversion

In the former chapters, we have discussed various relevant issues related to the measurement of real-world biosignals. These biosignals are typically of a continuous nature. This implies that at any point in time the amplitude has a defined value, and that the amplitude is a continuous function, as well. In nowadays measurements, we typically convert these *analog* signals to the digital domain, allowing e.g. additional digital signal processing, easy storage and retrieval. In this process, both time and amplitude are discretized, a process known as analog-to-digital (AD) conversion. In digitizing a signal, there are two essential elements involved: the sampling frequency and the AD-resolution. The sampling frequency is defined as the number of times per second a sample of the signal is evaluated. The higher the sampling frequency, the better the resolution is in the time domain. The choice of the appropriate sampling frequency is determined by the frequency content of the signal at hand. An important criterion in this respect is the Nyquist-Shannon sampling theorem, that states that the sampling frequency should be at least twice the highest frequency present in the signal [6, 7, 8]. This is illustrated in Figure 9.1. Here, we show a sine wave of 10 Hz, that is sampled with different sampling frequencies.

To satisfy the Nyquist-Shannon criterion, it is necessary to filter the analog signal, using a low-pass filter, before the actual AD conversion is realized. The cut-off frequency of this filter defines the highest frequency present in the

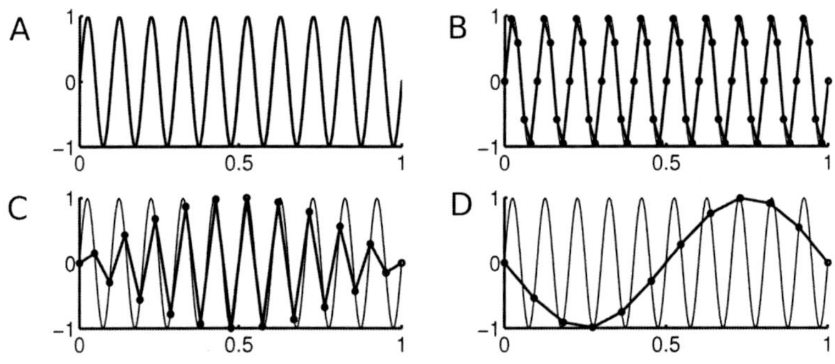

Fig. 9.1 Illustration of the aliasing effect. A sine wave of 10 Hz is sampled with a sampling frequency of 200 Hz (A), 50 Hz (B), 21 Hz (C) and 11 Hz (D). In panels B-D, the original sine wave is shown as a thin line. Using the sampling frequency of 11 Hz, aliasing is present: the frequency of 10 Hz now appears as a sine wave of 1 Hz. In other words: this frequency now presents as an other (incorrect) frequency, i.e 1 Hz, which explains why this phenomenon is known as aliasing (alias: assumed name, other name). Units on the x-axis are in seconds; y-axis shows amplitude in arbitrary units.

signal, based upon which the sampling frequency can be chosen, in practice typically about 3-5 times the cut-off frequency. These analog filters are also known as *anti-aliasing* or *pre-sampling* filters.

Analog filters gradually reduce the power of a signal above the cut-off frequency towards values near zero over a particular frequency range. The faster this is realized (expressed as dB attenuation per octave), the more powerful the filter is. This is one of the arguments why the sampling frequency should be chosen a factor 3-5 higher than the cut-off frequency of the analog filter. For EEG signals, for example, the analog filter is often set to about 70 Hz, and the sampling frequency is typically 250 or 500 Hz. Needle EMG recordings contain much higher frequencies that are clinically relevant, often up to 2 kHz. Therefore, in EMG recordings the sampling frequency is typically around 20 kHz, with analog filter settings around 3-5 kHz.

The resolution of an AD converter refers to the number of discrete steps with which the amplitude of the analog signal is displayed. In the sampling procedure, the amplitude of the analog signal is digitized into discrete levels, where the number of levels is a particular power of 2. This exponent is the number of bits of the AD converter. An 8-bit converter means that the amplitude can be discretized into $2^8 = 256$ levels. Since one of the bits is usually used as a sign bit the resolution is given by 2^{N-1} with N the number of bits of the AD-converter.

Most equipment for biosignal recordings use AD converters of 16 bit or more. The required AD-resolution is defined by the smallest amplitude that

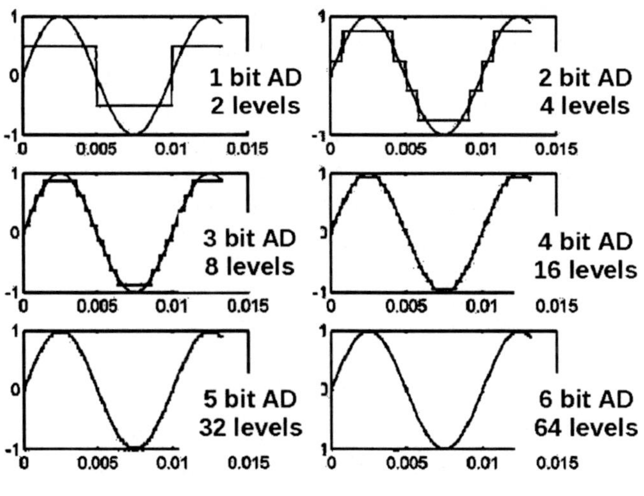

Fig. 9.2 Illustration of the AD-resolution for different AD-converters. At present, converters are typically 16-24 bits.

needs to be resolved and the range of the amplitude values. The effect of the AD-resolution on the displayed waveforms is illustrated in Figure 9.2.

After having sampled the data, data is typically stored to the hard-disk. Unfortunately, there are many data formats around, that makes sharing of data sometimes cumbersome. The European Data Format (EDF) and the recent extension to EDF+, proposed and developed by Bob Kemp and others [9], has definitely improved this situation, and many manufacturers now provide the possibility to transform their native data format to EDF or EDF+. For more information about EDF and EDF+, see `http://www.hsr.nl/edf`.

9.3 Fourier Series and Fourier Transform

The French mathematician Jean Baptiste Joseph Fourier (1768-1830) showed that it is possible to write any periodic time series $y(t)$ as a sum of (sometimes infinite) harmonic oscillations with discrete frequencies:

$$y(t) = a_0 + a_1 \sin(\omega_0 t + \phi_1) + a_2 \sin(2\omega_0 t + \phi_2) + a_3 \sin(3\omega_0 t + \phi_3) + \cdots \quad (9.1)$$

Here, ω_0 is the base frequency of the periodic signal The first coefficient, a_0 defines the DC component of the signal. This series is known as a Fourier Series. All sine waves are characterized by an amplitude (a_n), a frequency ($\omega_0 = 2\pi f_0$) and phase, ϕ. The number of terms can in principle be infinite. The amplitude, phase and frequency of all terms are defined by the signal $y(t)$. It is interesting to note that if we listen to a sound, the frequency we perceive is mainly defined by the lowest harmonic. The other harmonics define the color or timbre of the sound. Our ear is not very sensitive to phase differences. This implies that a signal that is composed of a sum of sinuses will appear the same if we add random phases to the different sinusoids.

Since a physical signal carries energy, we can also define a measure for the amount of energy that this signal dissipates. The various Fourier coefficients now represent these dissipations. The dissipation per unit time is "power". Therefore, we can write for the total power P_{tot}

$$P_{\text{tot}} = \sum_{i=0}^{N} a_i^2 \quad (9.2)$$

with N the total number of Fourier coefficients[1]. The squared value of an individual Fourier coefficient represents the contribution of the corresponding frequency to the total power of the signal. A graphical illustration of the squared value of these coefficients is therefore known as the (discrete) power spectrum. For completeness: a spectrum is a graphical representation of the

[1] Here, we have introduced a fictitious resistance of 1 Ω to deduct the power.

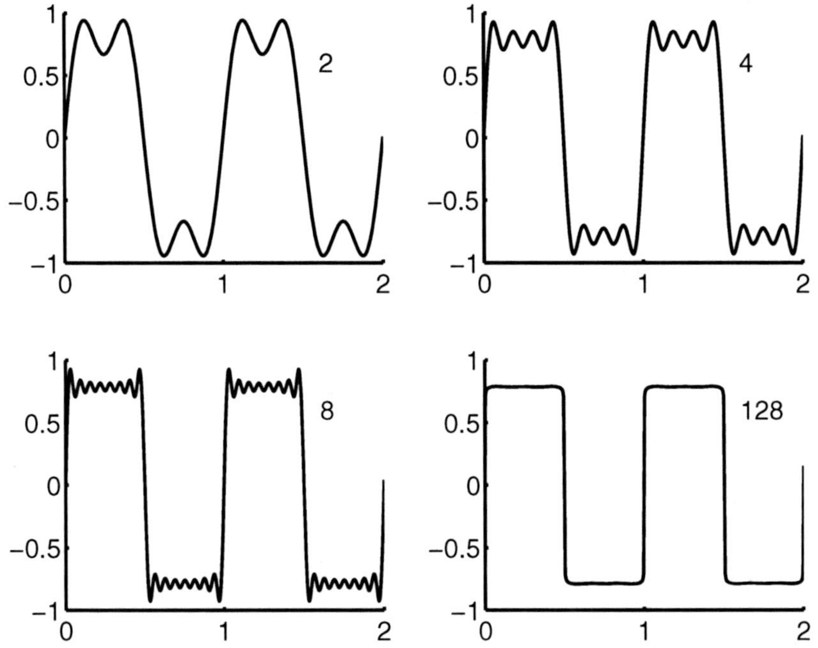

Fig. 9.3 Illustration of composing a square wave from an increasing number of sinus waves, odd multiples of the base frequency. The number of sinusoids increases from N=2, N=4, N=8 to N=128, respectively. The x-axis shows time; the y-axis amplitude in arbitrary units.

distribution of particular signal features (amplitude or power), often as a function of frequency (or wavelength).

We can create any periodic signal with a suitable combination of the various Fourier coefficients. An example is a square wave, as illustrated in Fig. 9.3. Here, the square wave is composed from an increasing number of sinuses, odd harmonics of the base frequency, ω_0, given by

$$y(t) = \sum_{k=1}^{N} \frac{1}{2k-1} \sin([2k-1]\omega_0 t). \tag{9.3}$$

In most cases, however, the signals we are faced with are not strictly periodic. In those circumstances, we cannot represent the signal as a series of Fourier frequencies. However, it is still possible to describe the signal with sine and cosine waves, but now with a continuum of frequencies. The sum will then need to be replaced by an integral sign. This is known as the Fourier Transform. An example of such a signal is a stationary, stochastic, non-periodic process: white noise. The spectrum of this signal is represented by a continuous horizontal line in the spectrum, indicating the presence of all frequencies, with the same power.

9.4 Fourier Analysis

Frequency analysis is an important and widely used technique in the quantitative analysis of the EEG. The goal is to extract clinically relevant information by decomposing the signal in its elementary building blocks: sine and cosine waves[2]. This process is known as Fourier Analysis or Spectral Analysis. This is complementary to the process described in the previous section. Now we start with the time series $y(t)$, and the goal is to determine the elementary frequencies. Mathematically, the problem can be stated as follows, were we will assume that the process generates a discrete spectrum. Therefore, we write

$$y(t) = \sum_{i=1}^{K} A_i \cos(\omega_i t) + B_i \sin(\omega_i t) \tag{9.4}$$

and the goal of the Fourier analysis is the find the correct values of the various unknowns, i.e. K, the coefficients A_i, B_i $(i = 1, \cdots, K)$ and the frequencies, ω_i $(i = 1, \cdots, K)$. This is the usual model for a process with a purely discrete spectrum (harmonic model) [8]. The constants A_i are defined as

$$A_i = \frac{2}{N} \sum_{t=1}^{N} X(t) \cos(\omega_i t), \tag{9.5}$$

$$B_i = \frac{2}{N} \sum_{t=1}^{N} X(t) \sin(\omega_i t). \tag{9.6}$$

Rewriting Eq. 9.4 to our previous expression,

$$y(t) = \sum_{i=1}^{K} a_i \sin(\omega_i t + \phi_i) \tag{9.7}$$

it holds for the parameter a_i that

$$a_i = \sqrt{A_i^2 + B_i^2}. \tag{9.8}$$

In practice, these parameters are typically *unknown*. This means, that for finding the various frequencies, we need to iterate the Eqs. 9.5-9.6 for all values of ω_i, in the hope that we will find non-zero values of A_i or B_i. In fact, we estimate the correlation between all "basic signal forms" (sinusoids) and the signal.

---

[2] Series of harmonic oscillations are sometimes also displayed as a sum of sine waves, with a phase angle ϕ. We will use both notations.

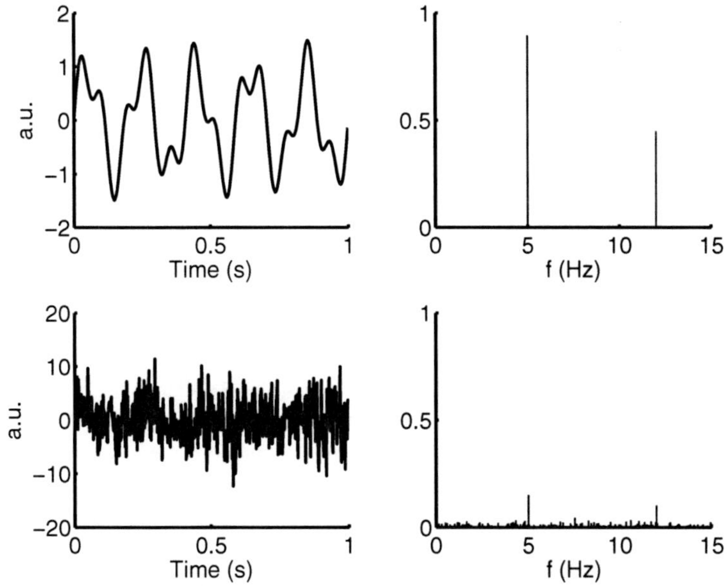

Fig. 9.4 Left the signal $y(t)$ as the sum of two sinusoids with frequency 5 and 12 Hz; some noise is added, as well. Right: the spectrum. Bottom left: the same signal, but with a much lower signal-to-noise ratio. The sine waves cannot be discerned now. In the spectral estimate, however, the two sinusoids are well visible at 5 and 12 Hz. This example illustrates the power of the Fourier Analysis very well: even the smallest sinusoid with a definite frequency can be found if "buried" in random noise.

As an example, consider a signal that is made from a sinus of $f_1 = 5$ Hz, and a sinus of $f_2 = 12$ Hz, plus a noise term, $n(t)$.

$$y(t) = \sin(2\pi f_1 t + \phi_1) + \sin(2\pi f_2 t + \phi_2) + n(t) \tag{9.9}$$

We will set $\phi_i = \phi_2 = 0$. Let us now multiply this signal $y(t)$ with sinuses of increasing and discrete frequencies e.g. $0.1, 0.2 \cdots, 1, 1.1, \cdots 50$ Hz. These are our essential waveforms. For almost all multiplications, we will find a correlation near zero. However, if we multiply our signal with a sinus of 5 or 12 Hz, the correlation will be significant and much higher than the other correlations. This is shown in Fig. 9.4. In fact, this procedure describes an elementary frequency analysis.

This search technique is in principle very time consuming. For instance, one assumes the presence of a particular frequency ω_i and estimates the value of the coefficient A_1 using Eq. 9.5. If the guess was right, and this frequency was indeed present in the signal, the value of A_i will be significantly larger than zero. If the frequency was not present, the value of A_i will be close to zero. By using a sufficient number of frequencies with sufficient resolution, and to repeat this procedure, we obtain a collection of amplitudes belonging to all the frequencies considered. This is the periodogram. In fact, the name periodogram is not well chosen, since it is a function of frequency and not period. The name was originally introduced by statisticians, who were looking for seasonal trends in data.

For processes with a continuous spectrum, the periodogram is a very bad estimate of the true spectrum. This is caused by the fact that a continuous spectrum (spectral density function) is approximated by estimating values of *discrete* frequencies. If too much spectral resolution is "allowed", the estimate will show large variations between successive frequencies, while for a continuous spectrum there should be some correlation between subsequent frequencies, and the appearance should be quite smooth.

In practice, this problem can be circumvented by limiting the spectral resolution by using a window function, and subsequent averaging of various estimates of the spectrum. By this procedure, a periodic signal is constructed with the period T defined by the window length, and therefore, the spectral resolution $\Delta f = 1/T$. This procedure also limits the variance of the final estimate. This technique is known as Welch's method, illustrated in Figure 9.5.

A limitation of the use of a window function is that some spectral components are given a value that is not exactly equal to their real value: spectral smearing. This is caused by the fact that energy of the frequencies that are present in the signal are not necessarily equal to the frequencies defined by the method: we now limit the spectral resolution to integer multiples of the minimum frequency. This can be compared to the creation of a histogram. In making a histogram, one defines a particular bin width. If the bin width is small, the estimate will not be very smooth, but each bin will represent the actual value. Increasing the bin width will typically reduce the variation between subsequent values, but some bias is introduced because different values are assigned to the same bin width. This uncertainty relationship, given by bias $\times$ variance $=$ constant, is known as Grenander's uncertainty principle.

Finally, in windowing the original time-series, it is necessary to assure that the start and end points of subsequent epochs are equal (typically zero). This is realized by multiplying the epochs with a particular window function. These windows are symmetrical functions, that multiply the time series with a function starting with zeros, gradually increasing to values one, and ending (in the tail) with zeros. Several of these windows exist. Well-known examples are Hamming, Hann, Bartlett, Kaiser, Gaussian and Welch. If such a window is not used, peaks may arise in the spectral estimate that are due due to discontinuities between subsequent epochs. This is known as a ripple artifact. Depending on the choice of the window, this artifact is more or less suppressed.

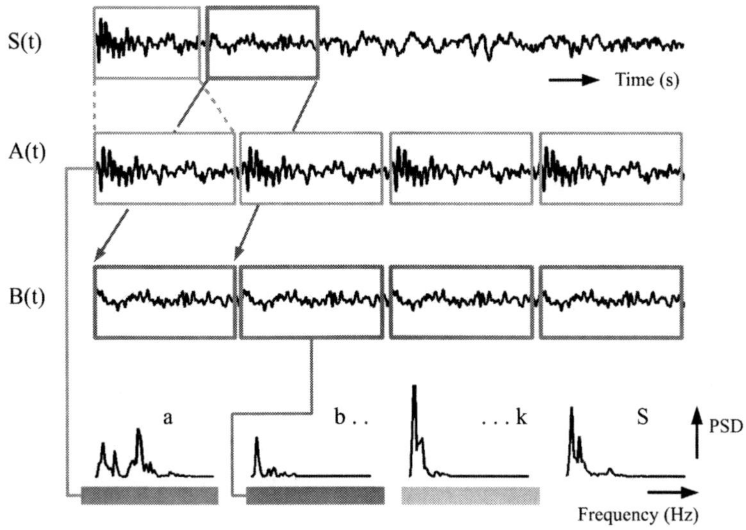

Fig. 9.5 Illustration of the frequency analysis as performed in the digital domain, using Welch's method. The original time series, S(t), is shown at the top. Two time series A(t) and B(t), composed of subsequent similar epochs, are illustrated below. The epochs shown are non-overlapping. For each of these time series, the spectrum is estimated, shown at the bottom. The final spectral estimate, S, is the average of the spectra $a, b, \cdots, k$. In practice, epochs are often half-overlapping. The periodic signal is constructed from a particular epoch using a window function, as well (not indicated) that assures continuity of the start and end points of the subsequent epochs.

Example

Consider a discrete signal, sampled with $f_s = 250$ Hz, given by

$$y(t) = \sin(2\pi \cdot 8 \cdot t) + \sin(2\pi \cdot 9 \cdot t) + \sin(2\pi \cdot 13 \cdot t) + \epsilon_t. \qquad (9.10)$$

This signal has a discrete spectrum, with spectral peaks of 8 Hz, 9 Hz and 13 Hz. Noise is modeled by ϵ_t. The desired spectral resolution is 1 Hz, the smallest difference between the three different oscillators.

If we would know the generator of this process beforehand, and if we would know that it generates a discrete spectrum, we could estimate the periodogram by using the whole signal length, N. In general, however, we do not have this prior information about the generator of the signal. Therefore, we often assume that the spectrum is continuous, and that we need to limit the spectral resolution, for reasons explained previously. We can now use Welch's method, and we will illustrate that this will indeed result in a *decrease* of the spectral resolution. Now, what is acceptable? The rule-of-thumb is that the window length need be longer than twice the minimal frequency resolution

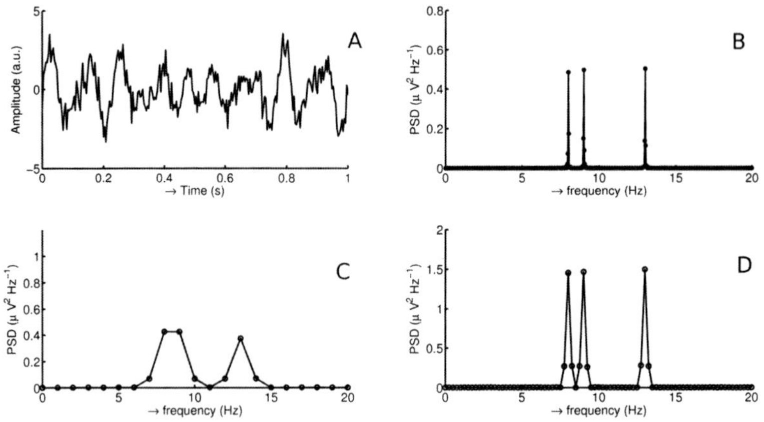

Fig. 9.6 Top left: time series, defined by Eq. 9.10. Three different spectral estimates from these time-series are shown, A-C. (B) Discrete spectrum estimated from the whole time series, without windowing. This shows a discrete spectrum, as generated by the model. (C): spectral estimate using an epoch length of 2 s (1024 samples) and using a Hamming window. The frequency resolution is now insufficient to resolve the 8 and 9 Hz, caused by this windowing procedure. (D) using a Hamming window with a length of 4 s. Now the spectral resolution is sufficient to resolve the frequencies of 8 and 9 Hz. Can you explain why the height of the spectral peaks in the various PSD plots are different?

that is desired. Assume, for our example, that we wish to differentiate spectral components that differ 1 Hz. We then need a window length larger than 2 seconds (i.e. 500 data points). Figure 9.6 presents the results of applying different window lengths, and the effect on the spectral resolution.

As a second example, we show an EEG signal (recorded at O2, common reference), sampled with $f_s = 500$ Hz, with length N=25000 (50 s data), as shown in Figure 9.7. If we create a periodogram with N=25000 (all data points), the frequency resolution is very high (1/50 s = 0.02 Hz), but the estimate of the true spectral density is not very reliable, since the variance is too large. If we use N = 2048, however, using a Hamming window, a more realistic estimate is obtained, as shown in the same figure (bottom right).

Another technique to reduce the variability of the spectral estimate is taking the mean of neighboring spectral components, using a "moving-average" filter. Such a filter calculates the running mean, by replacing each spectral component by a new value, calculated from the original value and the nearest neighbors, according to

$$A_{i,\text{smooth}} = \cdots b_{i-k}A_{i-k} + b_{i-k+1}A_{i-k+1} + b_iA_i + b_{i+1}A_{i+k+1} + \cdots \quad (9.11)$$

multiplying each value with a weighing factor b_i. In order not to change the amplitude of the spectrum the sum of the weighing factors should equal 1.

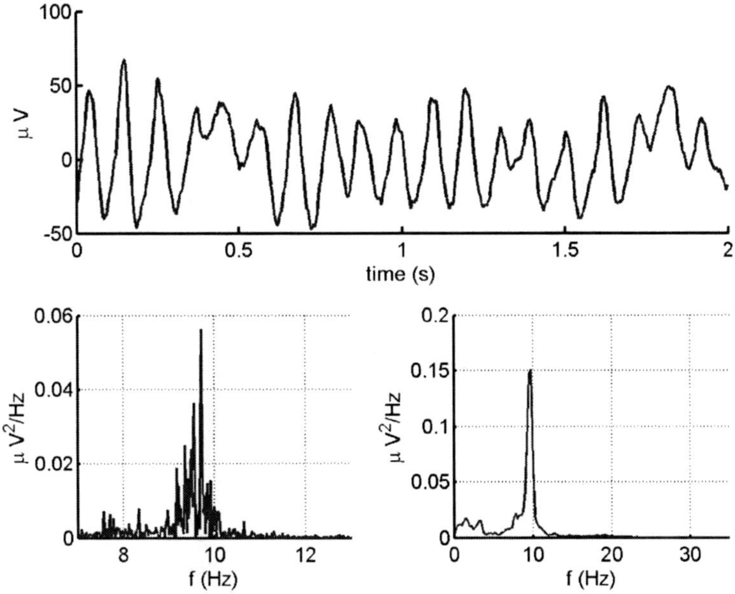

Fig. 9.7 Top: EEG signal recorded at O2 (common reference). Below two spectra. At the left the "raw" periodogram. Note the high variance. At the right the results of a spectral estimate using half overlapping windows with a duration of 4 s (Welch's method). This is a more realistic estimate of the true spectrum. The dominant frequency is approximately 10 Hz.

This procedure introduces a dependence of nearby frequency components, reducing the variance and increasing the bias, according to Grenander's uncertainty principle. Due, however, to the fact that the true continuous spectrum indeed will have correlations between neighboring frequency components, this procedure improves the estimate of the true spectrum.

It is also possible to estimate the spectrum by taking the Fourier transform of the autocovariance function[3]. In fact, this was the standard procedure before 1965. A disadvantage of this technique was that the process was (often) extremely time-consuming. The Fast Fourier Transform (FFT), introduced by Cooley en Tukey (1965) was a revolution, speeding up the calculation several orders of magnitude [8].

[3] The autocovariance represents the covariance between the data of the signal obtained at time t and the same signal, shifted in time (the "lag"). This function can be normalized; it is then known as the autocorrelation, with values between -1 and +1. The autocovariance or autocorrelation is therefore a function of the the time interval between signal pairs, and provides a measure for the similarity for each time lag. For most signals, this similarity decreases with increasing time shift. Signals that do not show this phenomenon remain correlated (they have memory), as,- in the trivial case-, holds for a strictly periodic signal, such as a sine wave.

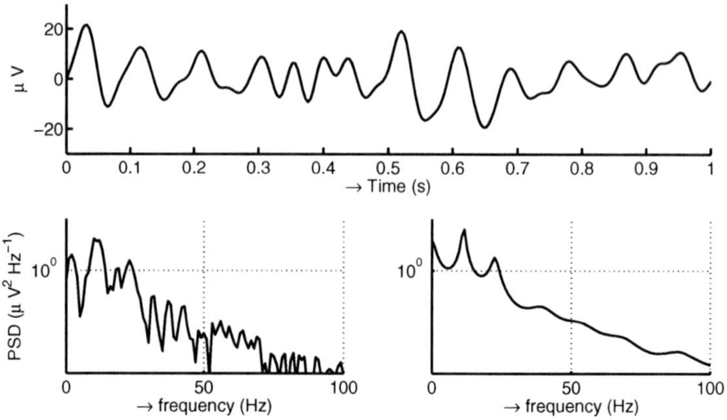

Fig. 9.8 EEG spectrum of 250 data points (1 second), filtered between 0.5 and 30 Hz. The time series is a bipolar recording in a healthy volunteer (P4-O2, eyes closed). The lower left panel shows the spectrum (on a semilog scale) as estimated with an FFT. Due to the short time series that is available, the estimate of the true spectrum is not very reliable. The lower right panel shows the estimated spectrum using an autoregressive model, calculated with Matlab routines (ARMASA) from [11]. This book discusses many more practical aspects of spectral analysis, and is a must-read for students with an interest in signal analysis.

Finally, there are other techniques that can be applied to estimate a spectral density function. e.g. using an autoregressive model. The spectral technique based on the Fourier Transform is a non-parametric technique, since no prior assumptions about a particular model for the time series are made. If we do have prior knowledge, for instance that the time-series can be modeled by an autoregressive (AR) model, it is possible to estimate the parameters of this model, that can subsequently be used for the spectral estimate. The advantage of the AR based spectral estimate is the frequency resolution, which is essentially unlimited, since the AR spectrum is a continuous function of frequency. In addition, if the time-series we have available is relatively short, using an AR based estimate may still provide us with a fair estimate of the spectrum[4]. This is illustrated in Figure 9.8). For more details about autoregressive spectral estimates, we refer to [8, 10, 11].

9.4.1 Practical Issues in Spectral Estimates

In practice, we deal with digitized signals. It is important to realize that the spectrum we obtain is an estimate of the true spectrum, often using a

[4] There are several techniques to determine the order of the AR-model; this is not always a trivial procedure; algorithms include the Levinson-Durbin and Burg algorithm.

Table 9.1 General guidelines for the choice of the epoch length and signal duration for a sufficient reliable estimate of the spectrum given a particular frequency resolution.

f_s (Hz)	Δf(Hz)	W_L (s)	N_{epoch}	signal duration (s)
250	1	2	>4	>8
250	0.5	4	>16	>64
500	1	2	>4	>8
500	0.5	4	>16	>64

Fast Fourier Transform (FFT) for our estimate[5]. There are two phenomena that can contribute to the difference in the estimated spectrum using the FFT and the true spectrum. First, aliasing. This can be reduced by proper analog filtering and choosing the correct sampling frequency. The second contribution is related to the windowing of the signal.

For a discrete signal with finite length and number of data points N, the Fourier series is not infinite, but contains a maximum of $N/2$ different oscillations. This is easy to check: say we sample with f_s Hz, and the signal duration is T s. The number of data points $N = f_s{\cdot}T$. Spectral analysis has as the highest frequency $f_s/2$ (Nyquist-Shannon) with frequency resolution $1/T$ Hz. Then it follows that the number of spectral components is $f_s/2{\cdot}T = N/2$.

The result of a Fourier analysis can therefore result in a frequency spectrum, with a value for all $N/2$ frequencies. In practice, we will typically not estimate all $N/2$ frequencies, but a fraction, due to the use of a window function with length M with $M \ll N$.

The motivation was mentioned previously: the estimate of a continuous spectrum is highly unreliable (high bias) if one uses the raw periodogram. The choice of the window function, including the window length, is defined by the spectral resolution and the desired bias.

Note, that due to the sampling, the spectrum obtained will always be discrete, even in those circumstances where the real spectrum is continuous. Therefore, the *effective* spectral resolution is in practice larger than 1/epoch length, since the use of a window function reduces the effective spectral resolution. This is due to spectral smearing. For instance, to estimate the power spectrum of the EEG alpha rhythm, we could use 20 s of EEG with 2 s windows. The *effective* spectral resolution is then slightly larger than 0.5 Hz, depending on the kind of window function used. In defining the spectral resolution, it is desired to differentiate different spectral peaks with sufficient reliability. It could be due to chance that spectral differences are observed. Similar to other measurements, it is necessary therefore to perform sufficient trials. To this end, epochs should be averaged, where the number needed is

[5] In fact, we obtain a Discrete Fourier Transform (DFT), using the FFT as an efficient algorithm.

given by [8]

$$N_{\text{epochs}} > \left(\frac{2}{\Delta f}\right)^2 . \tag{9.12}$$

Examples of the choice of the epoch length and signal duration for a desired effective resolution are presented in Table 9.1. Note that the sampling frequency is irrelevant for the spectral resolution. The maximum frequency in the spectrum is defined by the sampling frequency, as discussed previously.

Finally, we review some essential elements of the spectral analysis. If we perform a spectral analysis (typically using an FFT), and we sample with f_s Hz:

1. there is a signal amplitude every $1/f_s$ s;
2. the amplitude is discretized with a resolution dependent on the AD-converter;
3. a Discrete Fourier Transform (DFT) decomposes the signal into various sine and cosine waves with harmonic frequencies;
4. oftentimes, (non-)overlapping windows are employed with (W_L=window length in s), and the frequency components are multiples of $1/W_L$ Hz;
5. the highest frequency in the spectrum is $0.5 \cdot f_s$;
6. the actual frequency resolution is *larger* than $1/W_L$ because the use of a window function leads to spectral "smearing", reducing the effective resolution;
7. it is necessary to take the average of a sufficient amount of window epochs for a reliable estimate of the spectrum.

9.5 Clinical Applications of Digital EEG Analysis

To get a flavor of the use of quantitative EEG analysis, we will discuss shortly a few clinical situations in which feature extraction of the raw EEG signal is very useful.

9.5.1 *Intraoperative Brain Monitoring*

Patients with a stenosis of the carotid artery may suffer from Transient Ischaemic Attacks (TIA's), that are presumably caused by emboli, originating from the plaque that is present in this artery. Symptoms include a transient paresis or dysphasia. Often, in these patients an endarterectomy is performed. This is a surgical procedure in which the plaque is removed. The operation is typically performed under general anesthesia. In the course of the procedure, the artery needs to be clamped to allow the surgeon to remove the

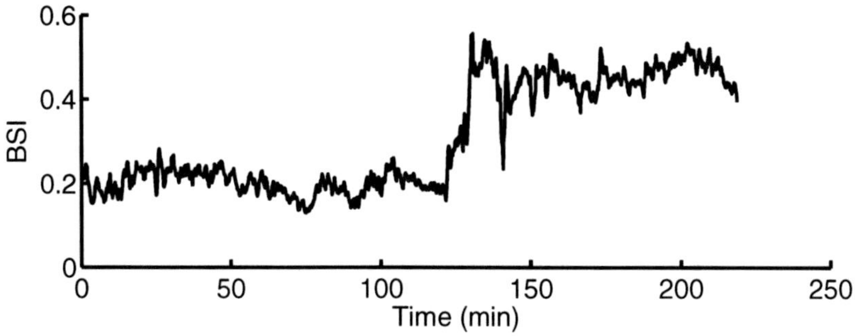

Fig. 9.9 Illustration of the applicability of the BSI for detecting focal seizure activity. Initially, there is only a slight asymmetry. After approximately 130 minutes from the beginning of the recording, a focal seizure starts, resulting in a significant increase of the BSI.

atherosclerotic lesion. At that moment, the circle of Willis[6] should allow sufficient blood to flow to this hemisphere. In a small fraction of patients, however, this is not sufficient for adequate cerebral perfusion, with the risk of a cerebral infarction. In most centers, the EEG is used to monitor the brain in these circumstances. As we discussed in the former chapter, if ischaemia occurs, EEG rhythms will change, as well. To assist in the visual interpretation, a technique was recently introduced to quantify this interhemispheric EEG asymmetry. The method presents a single number, the Brain Symmetry Index (BSI), as a measure for interhemispheric asymmetry. The BSI is defined as the mean of the absolute value of the difference in the mean spectrum of each hemisphere. The BSI is normalized, with BSI=0 for perfect symmetry and BSI=1 for maximal asymmetry. In physiological conditions, the BSI $\sim 0.06 - 0.08$. Although the technique was originally proposed to detect interhemispheric asymmetries during carotid surgery [12], the applicability of the BSI extends beyond detection of ischaemia [13]. Any change in spectral hemispheric symmetry can be detected with the BSI, including asymmetries due to focal seizure activity. This is exemplified in Figure 9.9.

9.5.2 *Visual Transform of Routine EEG*

Clinical EEG interpretation is still mainly performed by visual analysis, with a rather heuristic description of the various grapho-elements. This includes the evaluation of the spatial distribution of the various brain rhythms and the reaction to a variety of stimuli, including eyes opening and closing, hyperventilation, and photic stimulation. These features constitute the "background

[6] The circle of Willis is a circle of arteries, that creates redundancy in the cerebral circulation. If a part of the circle of Willis becomes compromised, blood flow from other vessels may preserve the cerebral perfusion. This includes the possibility to shunt blood from the internal carotid artery to flow to the opposite side of the brain.

pattern" of the EEG, because they define the mean statistical characteristics of the EEG signal, as was discussed in section 8.4 of Chapter 8, as well. In addition, the occurrence of various transients is noted, such as spikes, or spike-wave discharges. Besides the rather long learning curve associated with adequate visual EEG interpretation in a clinical environment, classic visual interpretation may suffer from various inter and intraobserver inconsistencies. In addition, it is rather subjective and cannot be used for continuous EEG monitoring, for instance in intensive care units. Furthermore, language is not

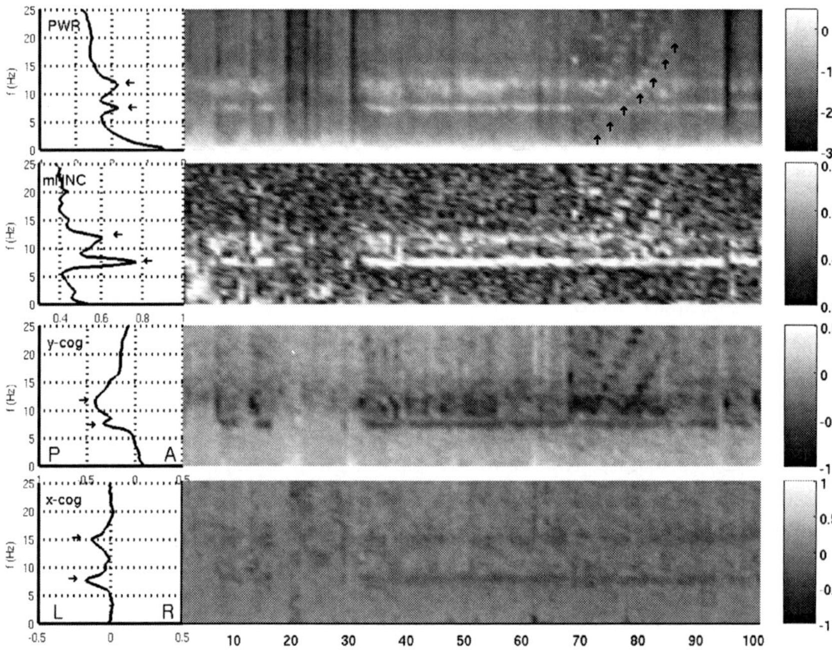

Fig. 9.10 Example of *The Colorful Brain*-transformation of a routine EEG. Four time frequency (TF) plots are shown, with time (in units of 10 s) on the x-axis and frequency on the y-axis. Mean values are displayed in the graphs on the left. The top panel shows the (normalized) power spectrum. There are two local maxima around 7.5 and 12 Hz (arrows). The next panel shows the maximum value of the mean coherence for neighboring electrodes (mNNC). There are two maxima, again around 7.5 and 12 Hz. The third panel shows the anterio-posterior distribution for each frequency. There is posterior (P) (black) dominance around 7.5 and 12 Hz. Finally, the bottom panel shows the left-right symmetry. There is a slight predominance of power at the left (L) side, in particular around 7.5 and 15 Hz. The parallel dotted lines seen in the upper three panels (arrows in top panel) result from photic stimulation, reflecting the induction of rhythmic activity by the stroboscope at the frequencies indicated. This is a normal EEG. Of course, the original illustration is in color. Various elements are preserved, however, in this black-and-white illustration.

always suited to communicate particular features. An example in this respect is the written report of a neuroradiologist. In almost all clinical situations, it is highly desirable to view the brain scan as well, to be adequately informed about the patients condition. Apparently, not all features allow translation into the language domain, and need to be presented to the clinician differently. This holds for EEG recordings, as well. Especially, elements of spatiotemporal dynamics are often difficult to "transform" into language. Alternative presentations, therefore, may assist in communicating about essential features, either physiological or pathological.

These considerations motivated the development of an alternative visual representation of the EEG, especially "designed" for applicability in a routine clinical environment. The transform, known as *The Colorful Brain* [14], is based on four[7] clinically relevant features of the EEG: spectral power, synchronization, and the left-right and anterio-posterior spatial distribution of the rhythms. These four features reflect important information about the mean statistical properties of the EEG (the "background pattern"). In addition, the transform is very compact, allowing to evaluate 20-30 minutes of a routine EEG in a single illustration. An example of this transform applied to a normal EEG is shown in Figure 9.10. For more details, see [14].

9.6 Summary

Biosignals are typically converted to digital signals into the electrical domain, to facilitate amplification, transport, filtering and subsequent signal analysis. We discussed various issues related to the sampling of analog signals and the analog-to-digital conversion. A short treatise on Fourier analysis and the estimation of a power spectrum has been discussed. Finally, this chapter presented some clinical examples of the use of quantitative EEG analysis.

Problems

9.1. What is the function of an analog filter for proper analog-to-digital signal conversion?

9.2. If you want to determine if the peak frequency between two time series differs by at least 1 Hz, how many epochs and what epoch length do you need for your Fourier analysis for a reliable estimate?

9.3. You want to resolve a signal with a resolution of 0.1 µV, and the amplitude varies between -100 tot +100 µV. What is the minimal requirement for the number of bits for your AD convertor?

[7] The original paper describes three time-frequency plots.

References

1. Northrop, R.B.: Signals and Systems Analysis in Biomedical Engineering. CRC Press, Boca Raton (2003)
2. Rangayyan, R.M.: Biomedical Signal Analysis. Wiley Interscience, Hoboken (2002)
3. Devasahayam: Signals and Systems in Biomedical Engineering. Kluwer Academic, Dordrecht (2000)
4. Niedermeyer, E., Lopes da Silva, F.: Electroencephalography: Basic principles, clinical applications and related fields, 4th edn. Lippincott Williams & Wilkins (1999)
5. van Drongelen, W.: Signal processing for neuroscientists: an introduction to the analysis of physiological signals. Academic Press, London (2007)
6. Nyquist, H.: Certain topics in telegraph transmission theory. Trans. AIEE 47, 617–644 (1928)
7. Shannon, C.E.: Communication in the presence of noise. Proc. Institute of Radio Engineers 37(1), 10–21 (1949)
8. Priestley, M.B.: Spectral Analysis and Time series. Academic Press, London (1981)
9. Kemp, B., Olivan, J.: European data format 'plus' (edf+), an edf alike standard format for the exchange of physiological data. Clin. Neurophysiol. 114(9), 1755–1761 (2003)
10. Thakor, N.V., Tong, S.: Advances in quantitative electroencephalogram analysis methods. Ann. Rev. Biomed. Eng. 6, 453–495 (2004)
11. Broersen, P.M.T.: Automatic autocorrelation and spectral analysis. Springer, Heidelberg (2006)
12. van Putten, M.J.A.M., Peters, J.M., Mulder, S.M., de Haas, J.A.M., Bruijninckx, C.M.A., Tavy, D.L.J.: A brain symmetry index (BSI) for online EEG monitoring in carotid endarterectomy. Clin. Neurophysiol. 115(5), 1189–1194 (2004)
13. van Putten, M.J.A.M.: The revised brain symmetry index. Clin. Neurophysiol. 118(11), 2362–2367 (2007)
14. van Putten, M.J.A.M.: The colorful brain: visualization of EEG background patterns. J. Clin. Neurophysiol. 25(2), 63–68 (2008)

Part III
Neurotechnology

Chapter 10
Neurotechnology

10.1 Introduction

The last decennium we have witnessed impressive advances in the development of technology to treat neurological dysfunctions. This includes various techniques to selectively stimulate abnormal functioning neural networks, thereby bringing their activity and function closer to the normal state. This chapter presents several new and exciting therapies, for example deep brain stimulation, vagus nerve stimulation for the treatment of some forms of epilepsy, brain computer interfaces and implants, such as the artificial retina and the cochlear implant.

10.2 Neurostimulation

Previously, we discussed techniques to stimulate nervous tissue in the peripheral nervous system, e.g. to estimate the motor nerve conduction velocity or the integrity of the spinothalamic pathways. As communication in the nervous system is mainly by electro-chemical signals, it is quite natural to explore if, in the case of abnormal electro-chemical interactions, electrical stimulation can restore particular functions to (near) normal. Both stimulation of the peripheral and the central nervous system is possible. For instance, in the peripheral nervous system, stimulation of the peroneal nerve is possible for the treatment of foot drop. Here, the primary goal of neurostimulation is to create a new (or boosted) signal in a peripheral nerve[1]. Engineering issues include selective stimulation, electrode design and the timing of the actual stimulus. A possibility is to have this controlled by a switch located under the heel bone of the patient: at the moment the heel bone touches the floor, the stimulator activates the peroneal nerve, and the foot is elevated. In this example, the working mechanism of the neurostimulation is well understood. In stimulation of an afferent nerve towards the brain, e.g. in the case when

[1] Direct stimulation of the muscle is possible, as well.

M.J.A.M. van Putten: Essentials of Neurophysiology, BIOMED, pp. 187–198.
springerlink.com

stimulation of the vagus nerve is performed for the treatment of some forms of epilepsy, the working mechanism is much less clear. This also holds for patients treated with neurostimulation in the central nervous system, either the spinal cord or the brain.

Electrical stimulation of the nervous system for therapeutic purposes dates back to the 18th century. In 1759, the Royal society published in the Philosophical Transactions a Letter from Doctor Himsel, at Riga, to Jacob de Castro Sarmento, MDFRS, that read[2]

> *On the 10th of March 1752, a young man 20 years of age, whose name is Mauve, in his infancy had a fall, by which his right arm became paralytic, of which he never had the least use from five years old. He was brought to me in order to attempt his cure by electricity; and he was in much the same condition with the patient cured by Mr. Jalabert at Geneva; who could not move his right hand in the least, and was afterward able to help himself, and follow his trade. All...... weakened. He approached the electrical tube, and touched it for some minutes; till the thenar, hypothenar, antithenar, indicator and the interossei, as well as the fingers of the lame hand, suffered electrical shocks one after another.*

If there really had been a long-lasting cure is not actually proven by this correspondence, but it does indicate that already more than 250 years ago, electricity was used in an attempt to treat neurological disorders. Interestingly, this "treatment" was communicated almost 30 years before Luigi Galvani published his findings on animal electricity in Italy, which took place in the year 1783.

Neurostimulation or neuromodulation refers to techniques in which electrical impulses are delivered to the central or peripheral nervous system, with the aim to cure or alleviate particular symptoms. At present, neurostimulation is used for the treatment of a variety of neurological disorders, including Parkinson's disease, tremor, Tourette's syndrome, chronic pain, epilepsy, depression and severe obesity[1].

10.2.1 Spinal Cord Stimulation

Spinal cord stimulation (SCS) is most commonly used in the treatment of chronic pain, that is insufficiently controlled by more standard treatments, including medication. SCS is currently used in patients suffering from peripheral neuropathies, root injury and phantom limb pain. Treatment effects differ, which is partially due to an insufficient understanding of the exact working mechanisms. This also applies to the underlying diseases that are treated with SCS, of which the pathophysiology is often not well understood, either. At present, this limits the efficacy of the treatment. It is hoped, that

---

[2] *The Case of a Paralytic Patient Cured by an Electrical Application*, enclosed in a Translated from the French Doctor Himsel and John Godfrey Teske, Philosophical Transactions (1683-1775), Vol. 51, (1759 - 1760), pp. 179-18.

with increasing insights into the mechanisms responsible for the development of e.g. chronic pain and an improved understanding of SCS, better treatments can be realized.

10.2.2 Vagus Nerve Stimulation

The number of patients suffering from epilepsy is significant: about 1% of the population suffers from this disease, of which a third cannot be treated adequately with medication. Alternative approaches to treating these patients include surgery and electrical stimulation of various structures in both the central and peripheral nervous system.

In the peripheral nervous system, stimulation of the vagus nerve is by now an almost standard treatment in particular patients suffering from drug-resistant seizures. The first application of vagus nerve stimulation (VNS) in humans dates back to 1988, when Penry implanted a VNS device in a patient with drug-resistant epilepsy [2, 3]. Soon after, this was followed by more human implants. The technique received a European CE mark approval for treatment in 1994. The FDA approved vagus nerve stimulation in 1997 as a clinical treatment for medically refractory partial epilepsy. During the surgical procedure, an electrode is placed around the left vagus nerve, that is connected to the stimulator by a subcutaneous lead. This pulse-generator is implanted under the skin of the left chest. An example of a vagus nerve stimulator (Cyberonics) is shown in Figure 10.1.

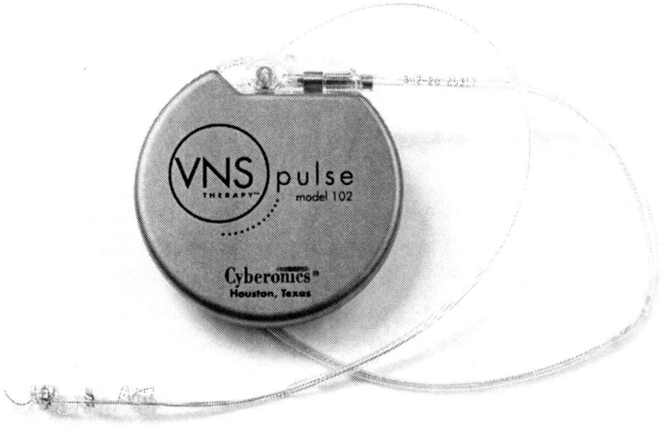

Fig. 10.1 The Cyberonics vagus nerve stimulator. The cuff, at the end of the electrode wire, is placed around the vagus nerve. Image kindly provided by Cyberonics (© Cyberonics), published with permission of Cyberonics, Inc.

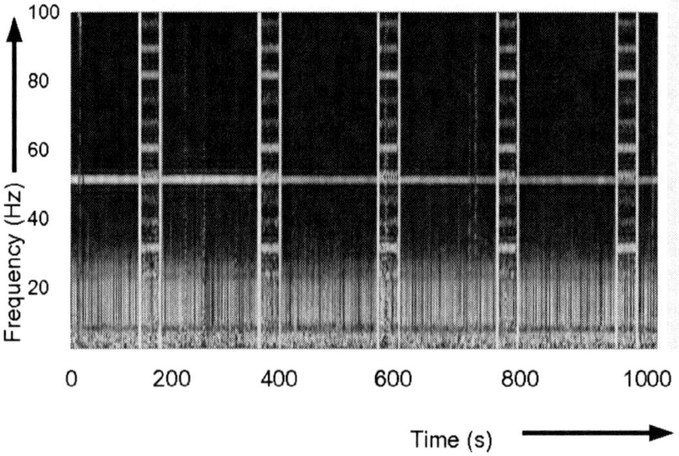

Fig. 10.2 Duty cycle of the vagus nerve stimulator in one of our patients. During the time that the stimulator is on (pacing at approximately 30 Hz), the time-frequency spectrum shows this and the higher order harmonics. The 50 Hz power line artifact is present, as well. Illustration slightly adapted from the MSc Thesis of Jelmer van Schoonhoven, University of Twente, Netherlands.

Although various stimulation protocols are used, a standard setting is that the device stimulates at a frequency of 30 Hz during 30 s, with pauses of about 2-3 minutes. This stimulation protocol is illustrated in Figure 10.2.

At present, VNS is used to treat adult patients with partial onset seizures. Initially, there is approximately a 30% reduction in the frequency of the partial onset seizures. After 2-3 years of treatment, the seizure frequency drops further, to 40-45%. In children with refractory epilepsy, VNS is used as well, with similar effects as in adults. In children suffering from the Lennox-Gastaut syndrome VNS has been applied, too. Here, seizure reduction rates vary from 27% to 64%.

The vagus nerve is composed of various fibers. The most abundant fibers (80%) are sensory C-fibers. These are unmyelinated and have the smallest diameter, and thus the highest excitation threshold when applying neurostimulation. Besides C-fibers, the vagus nerve contains myelinated B and A-fibers. The efferent fibers innervate the larynx and provide parasympathetic control to the heart, lungs and abdominal viscera. The right vagus nerve innervates the sinoatrial node of the heart, while the left innervates the atrioventricular node. It was shown in animal experiments (dog) that stimulation of the right vagus nerve caused more cardiac slowing that on the left. For this reason, in patients the left vagus nerve is used for stimulation. Which fibers are most relevant to stimulate, and what the optimal stimulation protocol needs to be, is not known in detail. The non-selective stimulation of the vagus nerve is responsible for the majority of side effects, that may occur. These include (in decreasing frequency) hoarseness, cough, pharyngitis, dyspnea, nausea, headaches and paresthesias. In many patients, these side effects nearly completely disappear after 1-3 years of therapy. For additional reading, see e.g. [4].

10.2.3 Deep Brain Stimulation

Although already in 1870 it was shown that electrical stimulation of the motor cortex in dogs can elicit limb movement, it is only recently that electrical stimulation of particular brain regions is applied at large scale to treat a variety of neurological and psychiatric disorders, including chronic pain, Parkinson's disease, tremor, dystonia and depression. In this technique, collectively referred to as deep brain stimulation (DBS), an electrode is placed in the brain, that modulates particular neural rhythms by sending electrical pulses to specific brain areas. Despite its long history, the working mechanisms of deep brain stimulation are still not well understood [5]. Traditionally, the belief is that therapeutic stimulation inhibits or excites neurons, where the modulation is a complicated function of the exact position of the electrode in the surrounding tissue and the stimulus parameters (current applied, stimulation frequency, duty cycle). DBS affects multiple neural elements, including myelinated axons and, to a lesser degree, cell bodies. In animal studies, it is also observed that various changes occur in the release of neurotransmitters or ATP. Stimulation of the subthalamic nucleus (STN) increases extracellular glutamate levels in the STN, globus pallidus interna (GPi) and the pars reticulata of the substantia nigra.

In the actual surgical procedure, an electrode is placed in the target region, using local anesthesia to allow feedback from the patient. The subsequent placement of the pulse generator occurs under general anesthesia. For Parkinson's disease, at present the most common side for electrode placement is the STN, GPi, or the brain stem pedunculopontine nucleus (PPN). In Figure 10.3 an X-Ray of the skull of a patient with a bilateral DBS electrode is shown.

Unfortunately, various side effects of DBS have been reported in the literature, as well. These include apathy, compulsive gambling and severe depression. Dysarthria and hypophonia are often reported in patients treated with subthalamic DBS. These side effects may occur in up to 9% of treated patients, and may be temporary. In some cases, the side effects are attributed to an incorrect calibration or placement of the electrode.

The present technology in DBS equipment is to a certain extent comparable to the early cardiac pacemakers. DBS is typically open-loop. Based on clinical examination of the patient and his/her reaction to the stimulation, the (technical) physician may decide to change particular stimulation settings. By recording signals from the electrode, DBS may become much more sophisticated, allowing closed-loop control, customized to the DBS target. Of course, a more profound understanding of the meaning of the various rhythms that we can record in the various brain structures and the relation to function

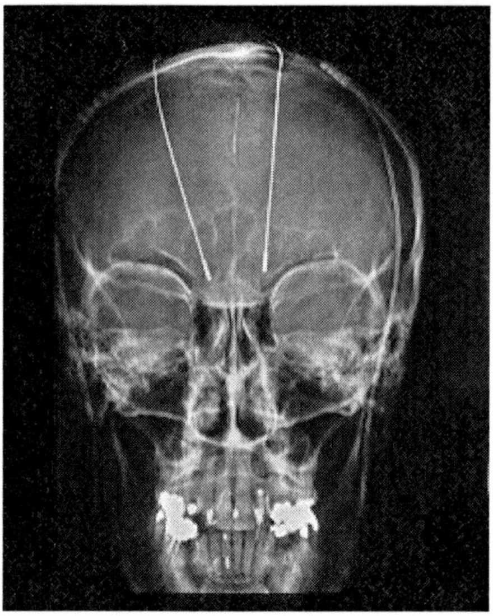

Fig. 10.3 Skull X-ray of a patient in whom bilateral electrodes are placed. The wire that connects the electrodes to the stimulator runs on the left side of the head. Picture kindly provided by Medtronic.

is necessary for these advances. Here, not only animal and human studies will contribute, but it is also foreseen that new mechanism will be revealed by the use of modeling and computer simulations.

10.3 Replacing Human Senses

The cochlear implant is - at present - the most successful implant to (partially) restore the function of one of the human sensors, needed for hearing. Hearing impairment is most often caused by a gradual loss of hair cells and nerve fibers. Because the nerve fibers without their hair cells will not be excited by the fluid displacement in response to the movements of the stapes, the spike activity in the auditory nerve will be a distorted version of the activity as produced by a normal hearing system. In these situations where the primary cause is degeneration of hair cells, conventional hearing aids will not be useful for the patient. Cochlear implants (CI), however, may be beneficial. A CI consists of an external microphone and a sound processor, that transmits the sound waves to a receiver and stimulator unit that is implanted in the bone above the patient's ear. The electrical stimuli are delivered to electrodes that are implanted in the cochlea, here stimulating the auditory nerve. Low-pitch tones will stimulate the longer, apical electrodes, while high-pitched tones will stimulate the shorter, more basal ones. This is similar to physiological sound processing, where the high frequencies are detected at the base and low frequencies at the apex. Of course, speech is a much more complex stimulus, that will be distributed over the various electrodes in the cochlea. Although the quality of hearing with a CI is far from normal, it is possible to understand spoken language sufficiently well to allow communication. In some patients, spoken language can even be understood without visual information from the movement of the facial musculature, in particular the tongue and mouth, allowing conversation by e.g. the telephone. Research to further improve CIs includes the use of directional microphone systems [6], dual electrode stimulation [7] and the effects of CIs on the quality of life [8]. Recent developments include auditory brainstem implants. These implants directly stimulate the cochlear nucleus of the brainstem in patients who are deaf because of bilateral cochlear nerve dysfunction [9].

Another area of interest is the development of a visual prosthesis by electrically stimulating the neural tissue of blind patients. In principle, several potential stimulation sites exist: the retina, the optic nerve, and the visual cortex. The first human experiments were started in 1966 with Giles Brindley's experiments [10, 11]. He applied electrical stimuli to the visual cortex, using 180 cortical electrodes. He was able to induce some visual sensations in his patients, called "phosphenes". An image, however, could not be created and useful vision was not produced. Although cortical stimulation is a possibility, the detailed physiology of cortical visual processing is insufficiently understood. Moreover, very dense electrode arrays are needed for sufficient spatially selective stimulation. Therefore, in patients suffering from a blindness due to diseases as retinitis pigmentosa or macular degeneration, stimulation of the retina or optic nerve may give better results. Patients with implants have a crude level of visual perception, but no visual prosthesis has yet restored vision. Problems with implants are varied, and include surgical complications, the power supply and the long-term viability of the implant. For instance, stimulation of the optic nerve induces phosphenes throughout the visual field. Some selectivity could be induced by changing the stimulus strength and duration in a patient.

Photoreceptors can be replaced by the use of retinal implants. The artificial silicon retina, containing thousands of photodiodes and microelectrodes, is placed in the subretinal space, between the outer retina and the retinal pigment epithelium. These implants do not need an external energy source, using the incident light for stimulation. If stimulated, the electrodes activate sensory neurons in the retina [12, 13]. The devices have been evaluated in a few patients. Visual perception was improved in patients with retinitis pigmentosa [14].

The epiretinal prostheses are placed on the surface of the retina. The electrical signals are received from a camera, that is positioned outside the body. There have been a few clinical trials, and patients reported simple visual perception, as phosphenes. Recently, biocompatible carbon nanotubes have been proposed for retinal prostheses. Highly selective stimulation of neurons is one of the key issues that limits the success of the visual prosthesis. In fact, electrodes should preferably be within 30 micrometers of the retinal cells; for a visual acuity of 20/80 this should even not exceed 7 micrometers. To realize electrode arrays consisting of thousands of electrodes, that also allow selective stimulation without cross-talk is an immense challenge. In the development of these high-resolution retinal prosthesis researchers explore various techniques, including the use of penetrating carbon nanotubes [15, 16].

10.4 Brain Computer Interfaces

Normal physiological control is realized by intact connections from and towards the brain. Various neurological conditions exist, where these normal connections are not functioning properly, such as amyotrophic lateral sclerosis, brain stem infarction or some forms of multiple system atrophy. Since decades, attempts have been made to restore some of the function in these patients. In the most severe cases, where all output channels are destroyed, the brain itself may still be functioning normally. This state is known as "locked-in": the self is enclosed in the brain, where communication to the outside world is not possible anymore.

A brain computer interface is aimed to partially restore this function, and can be defined as one or more communication and control channels that do not depend on the brain's normal output channels of peripheral nerve and muscle. A schematic illustration of a BCI-setup is shown in Figure 10.4. In order to realize such a system, various challenges have to be overcome. These include the recording of the proper signals, extraction of the relevant features, and using algorithms that allow the features to control particular instruments, e.g. a prosthesis or a cursor on a computer screen. Various approaches are possible. A very elegant technique was introduced by Farwell and Donchin [17, 18], using event-related potentials. The basic idea is based on the observation that if you experience an input stimulus that is different

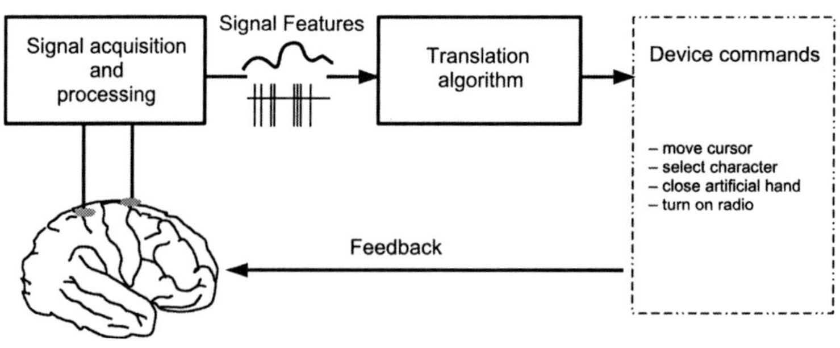

Fig. 10.4 Schematic of a BCI system and the necessary steps: data acquisition, feature extraction, control and output, with the feedback loop to the patient or volunteer.

from the standard input, about 300 ms after this "unexpected" stimulus, your brain generates a small local change in potential (the P300). In this BCI setup, the user is presented with a matrix of 6 x 6 cells, each containing one letter of the alphabet. To express a character, the user focuses attention on this particular character on the screen. At the same time, either rows or columns are intensified by a random generator. Every time when a row or column is intensified that contains the desired letter, a P300 is induced. By combining the row and column that elicits the response, it is known what character the user is intending to express. Although the system is rather slow (about one character per 5-10 s), it is actually being used by patients suffering from a locked-in condition.

Another technique to communicate is based on the possibility to modulate particular EEG rhythms at will. For instance, the μ-rhythm, that is generated over the motor cortex (positions C3 and C4), can be modulated by "thinking" about a movement, as we discussed in Chapter 8. By sufficient training and the use of real-time computer algorithms, this allows users to move e.g. a cursor across a computer screen. At present, these BCIs can realize bit rates of approximately 1-20 per minute. Various groups have shown that, by using the appropriate algorithms, these rhythms can indeed be used for elementary communication. With sufficient training, even 2-D cursor control has shown to be possible [19].

Invasive techniques are being explored, as well. The first successful realization used a special 100-microelectrode array, that was shown in Figure 5.8, implanted in the primary motor cortex of a tetraplegic patient (MN). From the thus obtained spiking activity, particular patterns could be extracted, that subsequently served as a control signal [20, 21]. This patient showed, using intended hand movements, that he could open his e-mail box or turn the television on, even while he was talking. He was also able to use this neural control to open and close a prosthetic hand. This study showed that neuromotor prosthesis, based upon the analysis of intracortical neuronal ensemble spiking activity, could eventually provide most valuable neurotechnology to assist in communication and control.

10.5 Neural Replacement

Although at present we can replace various organs, like the kidneys, the liver and the heart, replacement of damaged neural tissue to restore function is very limited. Cell replacement therapy for the treatment of neurological disorders, e.g. Parkinson's disease, spinal cord injury and motor neuronopathies, is a very challenging research area. Although cells of the central nervous system were once thought to be incapable of regeneration, in the last decade several studies have shown that migrating stem cells can be found in the brain. This was not only shown in animal models (e.g. in rodents, after inducing injury), but also in the human hippocampus [22]. Although at present, clinical applications are limited, in the next decade important contributions of these innovative therapies are foreseen [23].

Another, even more futuristic development, is the actual replacement of neural networks that are damaged by a stroke, trauma, or neurodegeneration. If the input-output relation of a particular neural network would be known in detail, this could perhaps eventually be realized by a silicon "brain-chip" or an in-vitro cultured neuronal network. This has the potential to treat some neurological disorders. Clearly, there are major obstacles here, including the interfacing with the living brain. In 2002, Theodore Burger and his team from the University of Southern California, US, started to test such a device in a slice preparation. After years of research, they were able to program the rat hippocampus input-output relations into a chip. As you recall, the hippocampus forms part of the limbic system, and has an important function in memory and navigation. These researchers removed a particular circuit in the hippocampus (CA3), and replaced this with their neural prosthesis (a Very Large Scale Integrated (VLSI) chip). By analysis of the temporal spike patterns in the slice, they claimed excellent agreement between the original activity, and the activity after replacement [24, 25]. Although this is only a first step, it shows a potential for restoring the function of neuronal networks, in this case the biological CA3 function.

It should be realized that this line of research is most likely still miles away from actual clinical applications. However, this is not always easy to foretell. The human mind is very creative, and advances in medical technology are often characterized by discontinuities, with unexpected breakthroughs. Of course, major ethical issues are involved in neural replacement therapies. Up to what point do we find it acceptable to influence the workings of the brain? Can it potentially lead to the control of functions, limiting the personal identity? Many people may find this technology hard to accept. At the same time, technology is very strongly affecting our lives, and this will only increase in the next decades. Technology has brought us many blessings, but - unfortunately - many disasters, as well. Here, scientists, engineers and physicians have a strong responsibility as to what to pursue and devote their careers to. In this sense, the future of biomedical technology and clinical engineering is also in your hands.

Problems

10.1. Review the anatomy of the vagus nerve, including the origin in the CNS. Note, that the vagus nerve both carries information from the brain as information towards the brain. Therefore, it is both an afferent and an efferent nerve.

10.2. What are cardiovascular effects of activity of the vagus?

10.3. The vagus nerve is composed of various fiber components, including A, B and C fibers. The diameters of these fiber types differ: A fibers are 5-20 μm in diameter, B fibers < 3 μm in diameter and C-fibers range from 0.4-2 μm in diameter. What does this imply for the conduction velocity?

10.4. Can you find some theories in the literature that are proposed to explain the working mechanism(s) of VNS?

10.5. How do you explain the main side effect of VNS, hoarseness? If you stimulate the VNS, what is the order of recruitment of the three different nerve fibers?

10.6. Can you make an estimate of the threshold (in mA) for the three different fibers?

References

1. DiLorenzo, D.J., Bronzino, J.D. (eds.): Neuroengineering. CRC Press, Boca Raton (2008)
2. Penry, J.K., Dean, J.C.: Prevention of intractable partial seizures by intermittent vagal stimulation in humans: preliminary results. Epilepsia 31(suppl. 2), S40–S43 (1990)
3. Uthman, B.M., Wilder, B.J., Penry, J.K., Dean, C., Ramsay, R.E., Reid, S.A., Hammond, E.J., Tarver, W.B., Wernicke, J.F.: Treatment of epilepsy by stimulation of the vagus nerve. Neurology 43(7), 1338–1345 (1993)
4. Lüders, H.O. (ed.): Deep Brain Stimulation and Epilepsy. Martin Dunitz. Taylor & Francis Group (2004)
5. Kringelbach, M.L., Jenkinson, N., Owen, S.L.F., Aziz, T.Z.: Translational principles of deep brain stimulation. Nat. Rev. Neurosci. 8(8), 623–635 (2007)
6. van der Beek, F.B., Soede, W., Frijns, J.H.M.: Evaluation of the benefit for cochlear implantees of two assistive directional microphone systems in an artificial diffuse noise situation. Ear. Hear. 28(1), 99–110 (2007)
7. Frijns, J., Kalkman, R., Vanpoucke, F., Bongers, J.S., Briaire, J.: Simultaneous and non-simultaneous dual electrode stimulation in cochlear implants: evidence for two neural response modalities. Acta Otolaryngol., 1–7 (December 2008)
8. Klop, W.M.C., Boermans, P.P.B.M., Ferrier, M.B., van den Hout, W.B., Stiggelbout, A.M., Frijns, J.H.M.: Clinical relevance of quality of life outcome in cochlear implantation in postlingually deafened adults. Otol. Neurotol. 29(5), 615–621 (2008)
9. Seki, Y., Samejima, N., Komatsuzaki, A.: Auditory brainstem implants: current state and future directions with special reference to the subtonsillar approach for implantation. Acta Neurochir. Suppl. 97(pt. 2), 431–435 (2007)
10. Brindley, G.S., Carpenter, R.H., Rushton, D.N.: Reaction times for simple shape discriminations requiring one or both visual cortices. Q. J. Exp. Psychol. 19(1), 70–72 (1967)
11. Brindley, G.S., Lewin, W.S.: The visual sensations produced by electrical stimulation of the medial occipital cortex. J. Physiol. 194(2), 54–55 (1968)
12. Ohta, J., Tokuda, T., Kagawa, K., Furumiya, T., Uehara, A., Terasawa, Y., Ozawa, M., Fujikado, T., Tano, Y.: Silicon lsi-based smart stimulators for retinal prosthesis. IEEE Eng. Med. Biol. Mag. 25(5), 47–59 (2006)

13. Zaghloul, K.A., Boahen, K.: A silicon retina that reproduces signals in the optic nerve. J. Neural Eng. 3(4), 257–267 (2006)
14. Chow, A.Y., Chow, V.Y., Packo, K.H., Pollack, J.S., Peyman, G.A., Schuchard, R.: The artificial silicon retina microchip for the treatment of vision loss from retinitis pigmentosa. Arch. Ophthalmol. 122(4), 460–469 (2004)
15. Gilmore, J.L., Yi, X., Quan, L., Kabanov, A.V.: Novel nanomaterials for clinical neuroscience. J. Neuroimmune Pharmacol. 3(2), 83–94 (2008)
16. Green, R.A., Williams, C.M., Lovell, N.H., Poole-Warren, L.A.: Novel neural interface for implant electrodes: improving electroactivity of polypyrrole through mwnt incorporation. J. Mater. Sci. Mater. Med. 19(4), 1625–1629 (2008)
17. Farwell, L.A., Donchin, E.: Talking off the top of your head: toward a mental prosthesis utilizing event-related brain potentials. Electroencephalogr. Clin. Neurophysiol. 70(6), 510–523 (1988)
18. Donchin, E., Spencer, K.M., Wijesinghe, R.: The mental prosthesis: assessing the speed of a P300-based brain-computer interface. IEEE Trans. Rehabil. Eng. 8(2), 174–179 (2000)
19. Wolpawand, J.R., McFarland, D.J.: Control of a two-dimensional movement signal by a noninvasive brain-computer interface in humans. Proc. Natl. Acad. Sci. USA 101(51), 17849–17854 (2004)
20. Donoghue, J.P.: Connecting cortex to machines: recent advances in brain interfaces. Nat. Neurosci. 5 (suppl.), 1085–1088 (2002)
21. Leigh, R., Hochberg, M.D., Serruya, G.M., Friehs, J.A., Hochberg, L.R., Serruya, M.D., Friehs, G.M., Mukand, J.A., Saleh, M., Caplan, A.H., Branner, A., Chen, D., Penn, R.D., Donoghue, J.P.: Neuronal ensemble control of prosthetic devices by a human with tetraplegia. Nature 442(7099), 164–171 (2006)
22. Hess, D.C., Borlongan, C.V.: Stem cells and neurological diseases. Cell. Prolif. 41(suppl. 1), 94–114 (2008)
23. Goldman, S.A., Windrem, M.S.: Cell replacement therapy in neurological disease. Philos. Trans. R. Soc. Lond B Biol. Sci. 361(1473), 1463–1475 (2006)
24. Hsiao, M.-C., Chan, C.-H., Srinivasan, V., Ahuja, A., Erinjippurath, G., Zanos, T.P., Gholmieh, G., Song, D., Wills, J.D., LaCoss, J., Courellis, S., Tanguay, A.R., Granacki, J.J., Marmarelis, V.Z., Berger, T.W.: Vlsi implementation of a nonlinear neuronal model: a "neural prosthesis" to restore hippocampal trisynaptic dynamics. In: Conf. Proc. IEEE Eng. Med. Biol. Soc., vol. 1, pp. 4396–4399 (2006)
25. Song, D., Chan, R.H.M., Marmarelis, V.Z., Hampson, R.E., Deadwyler, S.A., Berger, T.W.: Nonlinear dynamic modeling of spike train transformations for hippocampal-cortical prostheses. IEEE Trans. Biomed. Eng. 54(6 pt. 1), 1053–1066 (2007)

Part IV
Clinical Decision Making

Chapter 11
Rational Decision Making

11.1 Introduction

In clinical reality, decisions are made on an almost continuous basis. How will this patient react to this medication? How likely is it that this is the correct diagnosis? How reliable is this particular test? These are all clinically relevant questions that physicians and co-workers are faced with every day. It is important, therefore, to have some basic understanding of some concepts related to rational clinical decision making. This chapter will first present a general treatise on medical decision making. In the second part, test performance measures, such as the sensitivity and the specificity will be discussed. Subsequently, this chapter treats Likelihood ratio's, Bayes' formula, and presents various examples of the utility of various tests.

11.2 The "Monte Hall Three Door Problem"

To illustrate the relevance of additional information on a particular decision made, we will discuss a famous problem, "the Monte Hall three door problem" that is quite counter intuitive, but serves as a very good example of how chances can be modified by new data. The "Monte Hall three door problem" is an interesting, and generally puzzling, problem that has been around for years. In its most recent turn on the stage it was brought up by Marilyn Vos Savant in an article she wrote in the popular press (cited in [1]). The problem can be stated in the following way. Imagine that you are a participant in one of those television programs, where you have to answer more or less intelligent questions, perform a particular act, or show your tremendous memory capacity. And while doing so, you even reach the final. The quiz-master now shows you three doors on the stage, in front of you. He promises you that there is a car or a lot of money hidden behind one of the doors and he invites you to guess which door. If you guess correctly, you will win this price. If not, you go home with just "the honor". Suppose now that you guess door

M.J.A.M. van Putten: Essentials of Neurophysiology, BIOMED, pp. 201–208.
springerlink.com

Fig. 11.1 Cartoon of the Monte Hall three door problem. The candidate has chosen door B. The quiz master, however, has just shown that the price is not behind door A. Is the likelihood of winning the price changed now, and should you switch to door C?.

B. After your guess, however, the quiz master informs you that, of the two doors remaining, the car is not behind door A, which he proofs by opening this door. This situation is sketched in Figure 11.1.

Now the master of ceremony invites you to change your mind and go with door C rather than door B. Should you make the switch, should you stay with what you have chosen, or does it make any difference, anyway? This a basic illustration of decision making: Firstly, did we obtain new information by the opening of door 1, after the first decision was made (to go to door 2), and if so, how does this change the probabilities? Is it more likely now that the price is behind door 3, or are the chances even (50/50), or otherwise? The solution is left to you as an exercise. The ultimate answer will surprise many, and you should be very careful about the interpretation of the events that have taken place, and how this relates to the current situation. For further reading, see e.g. [2].

11.3 Changing Probabilities

The issues dealt with in this chapter are related to the fact that the number of *clinical presentations* of neurological diseases is much smaller than *the number of neurological diseases*. Therefore, patients with varying conditions may complain about loss of muscle strength; a patient who suffered from an episode where he/she lost consciousness may have suffered from a temporary disturbance in the blood pressure regulation or perhaps a seizure. Changes in sensation may be caused by various neuropathies. Headaches may have various causes, and pain in an extremity can be caused by a myriad of different diseases, as well. In general, history taking and neurological examination are not sufficient to come to a final diagnosis, and additional research is needed, for instance EMG, EEG, computed tomography or MRI. The question now naturally arises: how do the chances of a particular disease change by the test outcome? Note, that in clinical decision making we deal with diagnostic probabilities, i.e. the likelihood of a disease, *given* a particular symptom. In teaching, however, very often the emphasis is on *nosological* probabilities: *given* the diseases, what is the likelihood of a given symptom. The relationship between nosological and diagnostic probabilities is given by Bayes' formula, discussed in a later section. We will now proceed to a somewhat formal presentation of the decision process.

Let us first consider the collection of all possible diseases that a patient may suffer from. This collection will contain at least all diseases currently known, such as hepatitis, Alzheimer's disease and epilepsy. Let us call this collection V. Therefore, we write

$$V = (D_1, D_2, ..., D_n) \tag{11.1}$$

with D_i a particular disease and n the total number of possible diseases. The number n is very large, of course, given the enormous number of diseases we know today. In a particular clinical situation, however, this number will be smaller, but typically not equal to one. The fundamental problem in the diagnostic process is to estimate the probabilities of the conditions D_i for a given patient X

$$P_x(D_i), \tag{11.2}$$

for all the possibilities i in a particular clinical situation. We would now like to estimate all these probabilities for the patient X, where we want to find at least one D_i that is present with probability (near) 1. The actual probability that is accepted as sufficiently certain for a particular treatment (or sufficiently uncertain to not treat the patient) is a function of various variables, including the consequences of a wrong diagnosis, costs, delay of treatment, etc. The number of different conditions that is considered is in general a function of the clinical experience of the consulted physician and the complaint of the patient. This list of conditions D_i is the *differential diagnosis*.

In order to estimate the various $P_x(D_i)$ we need *information*. Several methods allow us to obtain this, ranging from history taking, the medical exam and additional investigations, like X-rays, CT-scans, electromyography, etc. The information that will be obtained from these investigations will be written as T_j, where the subscript j refers to the information element j. We will now state that T_j is diagnostically potentially relevant if

$$P(D_i \mid T_j) \neq P(D_i) \tag{11.3}$$

for at least one $i \in 1, 2, ...n$. Stated differently: Information is only diagnostically relevant if and only if the likelihood that a patient suffers from a particular condition changes by possessing this information. As an example, to ask a patient his favorite color for estimating the likelihood that he/she suffers from diabetes is irrelevant, i.e.

$$P(\text{diabetes} \mid \text{favorite color is red}) = P(\text{diabetes}) \tag{11.4}$$

while

$$P(\text{Alzheimer} \mid \text{does not know the current date}) > P(\text{Alzheimer}) \tag{11.5}$$

This process is known as differential diagnostic reasoning: the updating of probabilities by acquisition of new information. Clearly, the more reliable we consider a particular test, the more likely it is that, given a positive (negative) test outcome, the patient will (not) suffer from a disease. We will now discuss in more detail how this likelihood can be quantified, and how the prior likelihood and the test quality contribute to this probability.

11.3.1 Quantifying a Test Outcome

We are now faced with the question how to define the quality of a test outcome. As an example, consider a test that is able to predict if a patient may suffer from Alzheimer's disease. The test results are presented in the following table, Table 11.1. Is this a useful test? It depends. As stated in the previous section, the primary goal of a test is to obtain information, that changes the likelihood of the presence of a particular condition. In addition, it is relevant to be informed about the change in this likelihood, and its absolute value. For some conditions, it is acceptable to be not 100% sure about the diagnosis

Table 11.1 Example of possible test outcomes for a test applied to 100 patients. D+ is the presence of the disease, D- the absence. Test+ indicates a positive test outcome, Test- a negative test outcome.

	D+	D-
Test +	45	4
Test -	3	48
Total	48	52

(e.g. if treatment options are not possible in the case that the diagnosis is certain), whereas if a particular treatment is available that bears particular risks, one needs a high degree of confidence in the diagnosis.

To come back to our example, it is shown that our test is positive in most cases that the disease is present (45 times), but, unfortunately, our test fails to become positive 3 times. This implies that the *sensitivity* of our test, defined as P(T+|D+), equals $45/48 \sim 0.94$. In words: the sensitivity is the likelihood of obtaining a positive test outcome, given the presence of the condition. On the other hand, if the diagnosis is absent (D-), our test fails to recognize this in 4 out of 52 cases. In the remainder of cases, the test correctly identifies patients without the condition. This is defined as the *specificity*: the likelihood of a negative test outcome, given the absence of the condition. In this example, therefore, we find that the specificity, P(T-|D-)=$48/52 = 0.92$. Note, that sensitivity and specificity relate to the quality of the *test*, it is *not* a number that tells you how likely it is that a patient suffers from a particular condition. Of course, there is a particular relation between the test performance, as quantified by the sensitivity and specificity, and the likelihood of a disease. This relationship is discussed next.

11.3.2 Bayes

Every patient that is consulted by a particular physician in a certain setting (country, area, city), may suffer from a condition. This likelihood is also known as the a priori likelihood. As an example, if a patient complains about a fever, a headache, and on examination is found to have a stiff neck, the prior likelihood of meningitis is near 0.9. If on lumbar puncture, you find a high cell count, low glucose, and a Gram positive stain, your diagnosis of a bacterial meningitis is certain. This is your posterior likelihood, obtained after the test. In fact, you already performed a few other tests, i.e. the physical examination. If the patient has no fever, and tells you that he/she has been suffering from a stiff neck since a few weeks, the prior likelihood of meningitis is near 0. Oftentimes, however, the likelihood is somewhere in between, say 0.3 to 0.6. Bayes formulated a very useful relation between the a posterior likelihood $P(D \mid T)$ and the prior likelihood, $P(D)$, given by

$$P(D \mid T) = \frac{P(T \mid D)P(D)}{P(T)} \tag{11.6}$$

where T is the presence or absence of the diagnostic element, i.e. the test. Bayes' formula states the essence of the diagnostic process, that may be iterated if new information is obtained. In various circumstances the physician will not "calculate" the various a posterior probabilities for a single information element, but for a cluster of these elements. In fact, this is very similar to some pattern recognition techniques. Bayes' equation tells you, that the likelihood of a disease, given a test, is equal to the likelihood of a test outcome, given the disease, multiplied by the prior likelihood of the disease divided by the likelihood of the test outcome. Note, that this latter likelihood is based upon a test evaluation, regardless of having or not having the disease. We will now slightly rewrite this expression, to the situation where the test is positive, and we wish to estimate the likelihood of the disease. This results in:

$$P(D \mid T+) = \frac{P(D)}{P(D) + \frac{1-P(D)}{LR^{+}}} \tag{11.7}$$

with the positive likelihood ratio defined as

$$LR^{+} = \frac{P(T+ \mid D+)}{P(T+ \mid D-)} \tag{11.8}$$

with $D+$ the presence and $D-$ the absence of the disease.

In a similar vein, we can write the expressions for a negative test outcome. For the conditional probability we now have

$$P(D \mid T-) = \frac{P(D)}{P(D) + \frac{1-P(D)}{LR^{-}}} \tag{11.9}$$

with the negative likelihood ratio defined as

$$LR^{-} = \frac{P(T- \mid D+)}{P(T- \mid D-)}. \tag{11.10}$$

Therefore, two likelihood ratios exist, the positive likelihood ratio, LR^{+}, and the negative likelihood ratio, LR^{-}, for a positive and a negative test outcome, respectively. The positive and negative likelihood ratios can also be written as a function of the sensitivity and the specificity. We then obtain for the positive likelihood ratio

$$LR^{+} = \frac{\text{sensitivity}}{1 - \text{specificity}} \tag{11.11}$$

and for the negative likelihood ratio

$$LR^{-} = \frac{1 - \text{sensitivity}}{\text{specificity}}. \tag{11.12}$$

In Table 11.2 we show some values of the likelihood ratios for some tests. In practice, there is typically a trade-off in defining a test: increasing the sensitivity (e.g. by lowering a particular detection threshold) will *reduce* the specificity, and vice verse. This relationship can be expressed in a receiver operating characteristics (ROC) curve. The ROC curve has the sensitivity plotted vertically and 1-specificity on the horizontal axis. The sensitivity and specificity, and therefore the performance of test, vary with the cut-off. If several tests are performed on the same sample, the various

Table 11.2 Illustration of some Likelihood ratio's. For more data, see `http://www.cebm.utoronto.ca/glossary/lrs.htm#table2`. Data not available is denoted with n.a.

Disorder	**Test**	**Standard**	**Setting**	**LR$^+$**	**LR$^-$**
dementia	MMSE < 27	diagnosis of dementia by psychiater using DSM-III-R	University geriatric hospital in Switzerland	2.5	n.a.
adult meningitis	jolt accentuation of headache	LP or autopsy	headache and fever; confirmed meningitis	2.2	0
Alzheimer's Disease	Positive apolipoprotein E (APOE) genotype test	Criteria from the National Institute of Neurological Disorders and stroke or the DSM-III-R	Patients referred for evaluation of dementia in AD centers in US	2.0	n.a.
Carpal Tunnel Syndrome	Fingers involved	Abnormal motor or sensory conduction within the carpal tunnel, measured by nerve conduction testing	Patients presenting with upper extremity discomfort	2.4	0.5
dementia	clock drawing test	NINCDS-ADRDA criteria	elderly patients	9.6	0.4

ROC curves can be used to compare their performance. As directly follows from Bayes' equation, the final relevance of a test in a clinical situation is both a function of the test-quality, defined by the likelihood ratio, and the prior likelihood. This relationship can be expressed graphically, as well. In such a plot, the relationship between pre-test probability of a disease and the post-test probability is indicated, in relation with the likelihood ratio for the test applied. The student is referred to the exercises at the end of this chapter to further illustrate this point. This chapter only presents a few relevant topics of clinical decision making. There are many good books available that treat these and other issues in much more detail [3, 4, 5].

11.4 Summary

In the diagnostic process, prior knowledge and prior likelihoods can be changed by additional information. The amount of change is a function of the prior probability and the sensitivity and specificity of a particular test, that can be described by Bayes formula and the likelihood ratio.

Problems

11.1. What is the relation between the a posterior and a prior probabilities, given the sensitivity and specificity of a test? Make a graphical illustration of this relationship for different values of the likelihood ratio, e.g. LR^{+}=10 LR^{-}=0.1 and LR^{+}=1. Plot prior probability on the x-axis and posterior on the y-axis. Explain in which circumstances a test is most useful.

11.2. You estimate the prior probability of a disease to be 30%. You use a test with a positive likelihood ratio of 1. The test is positive.
a) What is the likelihood of the disease having the test outcome?
b) You use another test, with a sensitivity of 90% and a specificity of 90%. This test is positive, as well. What are the chances of the presence of the disease now?

11.3. What is the solution to the Monte Hall three door problem?

11.4. A test has a sensitivity of 80%. If, with an a prior probability of 30% and a positive test outcome, the a posterior probability must be larger than 98%, what is the positive likelihood ratio and the specificity of this test?

11.5. A patient suffered from a seizure. Your guess is that the chance that this is the first manifestation of epilepsy is approximately 50%. You record an EEG. The EEG is normal. Which of the following is true?

a) The likelihood of epilepsy is near zero, since the sensitivity and specificity of the EEG are around 98%
b) The likelihood of epilepsy is still high, 30-40%
c) The likelihood of epilepsy is almost nil since the specificity is very high
d) The likelihood of epilepsy is almost nil since the sensitivity is very high.

11.6. Assume that a particular test has a positive likelihood ratio $LR^{+} = 10.4$ and a negative likelihood ratio $LR^{-} = 0.07$. How good this is in practice, is strongly dependent on the a prior knowledge. Consider for instance a situation where the a prior likelihood to suffer from Alzheimer's disease is 0.80. Given a negative outcome of this test, what is the likelihood of Alzheimer's disease?

References

1. Bohl, A.H., Liberatore, M., Nydick, R.L.: A tale of two goats and a car, or the importance of assumptions in problem solutions. Journal of Recreational Mathematics, 1–9 (1995)
2. Bapeswara Rao, V.V., Bhaskara Rao, M.: A three-door game show and some of its variants. The Mathematical Scientist 17(2), 89–94 (1992)
3. Kreyszig, E.: Introductory Mathematical Statistics. John Wiley and Sons/Wiley, New York (1970)
4. Ingelfinger. Biostatistics in Clinical Medicine (1999)
5. Campbell, M.J., Machin, D.: Medical Statistics a commonsense approach. John Wiley and Sons, Chichester (1999)

A
Formula

Table A.1 Constants and formula

Nernst	$V_m = \frac{-KT}{q} ln \frac{[A^+]^i}{[A^-]^o}$ (A is a particular ion)	mV
length constant	$\lambda = \sqrt{\frac{ar_m}{2r_{ax}}}.$	m
current injection	$V_m = V_o e^{-\|x\|/\lambda}$	V
voltage clamp	$V = V_0 \left(1 - \exp^{-t/\tau}\right)$	V
ap conduction velocity	$v_{\text{ap}} \sim 2\lambda/\tau_m$	m/s
specific electrical resistance	$\rho = RA/l$	Ω m
electrical conductivity	$\sigma = 1/\rho$	S/m
current density	$J = \sigma E$	C/m^2
diffusion flux	$J = -D \frac{\partial \phi}{\partial x}$	$mol/(m^2s)$
Bayes	$P(D\|T) = \frac{P(D)}{P(D) + \frac{1-P(D)}{LR}}$	-
Likelihood ratio (LR^+)	$LR^+ = \frac{\text{sensitivity}}{\text{1-specificity}}$	-
Likelihood ratio (LR^-)	$LR^- = \frac{\text{1-sensitivity}}{\text{specificity}}$	- -
firing rate (IF neuron)	$f_r \approx \left[\frac{V_{\text{rest}} - V_{\text{th}} + R_m I_e}{\tau_m (V_{\text{th}} - V_{\text{reset}})}\right]$	Hz

B
Symbols and Physical Constants

Table B.1 Symbols and Physical constants

Symbol	**meaning**	**value and unit**
x	distance along axis	m
t	time	s
k	Boltzmann's constant	1.38 10^{-23} J/K
T	absolute temperature	Kelvin
q	electron charge	1.6 $\times 10^{-19}$ C
R	gas constant	8.31 J/(mol K)
F	Faraday's constant	96485 C/mol
D	diffusion coefficient	m^2 s^{-1}
N	Avogadro's number	6.02 10^{23} molecules/mol
E	electrical field strength	V/m
V	voltage	V
I	current	A
β	partition coefficient	-
ϵ_0	permittivity of free space	8.85 10^{-12}F m^{-1}
ϵ_r	relative (static) permittivity (ϵ_c/ϵ_0)	
ϵ_r	dielectric constant of myelin	7
r_m	specific membrane resistance	Ω m^2
	typical value	0.1-1 Ω m^2
c_m	specific membrane capacitance	F/m^2
	typical value	0.01 F/m^2
r_{ax}	specific axial resistance	Ωm
	typical value	0.1-1 Ωm
σ	specific conductivity ($1/r_{ax}$)	S/m
R_{ax}	axial resistance	Ω
R_m	membrane resistance	Ω
C_m	membrane capacitance	F
d_m	thickness of myelin	m
δ	width of node of Ranvier	1.5 $\times$ 10^{-6} m
a_1	inner radius of axon	0.5-10 $\times 10^{-6}$ m
a_2	outer radius of axon	1-15 $\times 10^{-6}$ m
L_r	distance between nodes of Ranvier	3-30 $\times 10^{-4}$ m

Solutions

Problems of Chapter 2

2.1 Consciously perceived stimuli are e.g. sound, vision, touch. Unconsciously, we perceive for instance blood pressure, sodium concentration in the blood plasma, pH. Many other examples exist. Note, that in some circumstances the "level of conscious perception" may change. Depending on focus, you may e.g. not be (consciously) aware of environmental noise.

2.2 C-fibers → brain (pain experience) → motor cortex → muscle → movement→ reduction of excitation of C-fibers.

2.3 see text.

2.4 Na, K. If you inject too much K^+ intravenously, extracellular K^+ will increase, and the resting membrane potentials will increase. Hyperkalaemia initially causes membrane excitability due to partial membrane depolarization. As the level of serum potassium increases the membrane will remain depolarized and action potential propagation is blocked. (Recall, that repolarization is necessary before a new action potential can be generated and transported).

Hyperkalaemia causes muscle weakness, arrhythmia's, and eventually cardiac arrest. The ECG changes are: peaked T waves, decreased P-waves, QRS widening, and sine wave arrest.

2.5 -20.8 mV. If you make a program in Matlab, using the GHK equation, you can easily get a feel for the effect of changing concentrations and permeabilities.

2.6 See text, figure 2.10. Ion gates are closed and opened by changes in transmembrane voltage.

2.7 Hyperpolarization is caused by a relative increase in permeability for potassium ions, as compared to the permeability for potassium in the resting condition. In addition, there is a transient de-activation of the sodium permeability (cf Figure 2.10). The action potential ends when both the permeability of the potassium and sodium return to normal resting levels.

2.8 If you have made your Matlab program, as suggested previously (Q5), you will find that these effects are very small. The reason is that during rest, the permeability for potassium is much higher than for sodium. Use e.g. the data from Table 2.3, lowering the intracellular sodium concentration from 12 to 5 mmol/l.

Evaluating at T=20 degrees C, we find for the values given that $V_m = -85.422$ mV. Lowering sodium to 5 mmol/l, we obtain $V_m - 85.417$ mV.

2.9 Using values from Table 2.3. With $K^+ = 150$ mmol/l V_m=-85.1 mV, with K^+=135 mmol/l, V_m=-84.1 mV.

2.10 Both diffusive and electrical forces now work in the same direction, since at this phase the membrane is negatively charged on the outside.

2.11 d.

2.12 b.

2.13 a.

2.14 c.

2.15 a) $A = 2\pi rl \sim 6 \times 10^{-6}$ m^2.
b) $C = (3)(8.85 \times 10^{-12} C^2/(Nm^2)) \frac{6\times10^{-6}}{10^{-8}} \sim 10^{-8} F$. Total voltage change is 100 mV. Using Q = CV we find that the total change in charge is 10^{-9}C, that approximates 10^{10} ions.
c) The volume of the axon piece is V=$\pi r^2 l \sim 3 \times 10^{-11} m^3$. This holds 3 $\times 10^{14}$ sodium ions. The relative change is $10^{10}/(3 \times 10^{14}) \sim 0.33 \times 10^{-4}$.

Problems of Chapter 3

3.1 See text.

3.2 See text.

3.3 See text.

3.4 See text.

3.5 (a) 13 mm; b) ~ 0.2 mm.

3.6 Approximately 13 m/s

3.7 Simply inserting the above expression in the cable equation proofs it is a possible solution. Therefore, the space constant determines the change in membrane voltage as a function of distance along the axon. Looking again at the expression for λ, Eqn 3.10, can you sketch an elementary electrical circuit that shows this relationship, as well?

3.8 Na-channels are extremely sensitive to temperature; as temperature rises, Na-channels open much more quickly in response to depolarization. As the cycle of Na-channels activity accelerates, the duration of the AP is reduced. A 2 degrees Celsius rise in temperature, such as can occur during exercise,

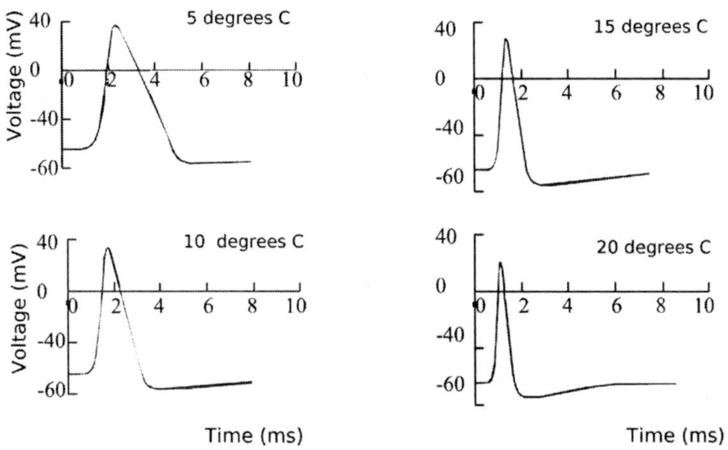

Fig. B.1 Generation of four action potentials at four different temperatures. Due to the strong temperature dependence of the voltage gated channels, the action potential duration, width and amplitude are strongly temperature dependent, as well. Simulations performed with Neuron.

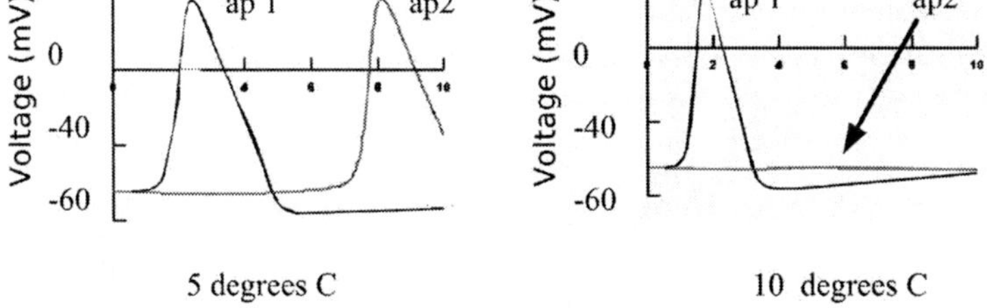

Fig. B.2 Generation of two action potentials at two different temperatures and registration of voltage changes along the axon (gray square). At 5 degrees C, the action potential is propagated along the axon towards the recording site; at 10 degrees C, however, action potential propagation is blocked. Simulations performed with Neuron.

can decrease the duration of the action potential by 20%. This has the effect of reducing the duration of the local depolarizing currents which are responsible for "carrying" the AP between Nodes of Ranvier. Normally, this has little effect; AP's are still propagated between Nodes of Ranvier. However, in multiple sclerosis, lesions in myelin can so compromise the propagation process that the reduced duration of the local depolarizing currents can result in a failure of propagation that had been occurring (albeit, just barely) at the lower temperature. Significantly, this exacerbation of the disease process is immediately relieved in patients by cooling. Results using NEURON (`http://neuron.duke.edu/`) to simulate the effect of temperature on the action potential shape are shown in Figure B1 and Figure B2.

3.9 In a severe motor neuropathy you can find muscle weakness, low or absent reflexes, preservation of sensory functions and fasciculations. Guillain-Barré can develop in a few days, and is an autoimmune disorder.

Problems of Chapter 4

4.1

Table B.2 Some differences between an electrical and a chemical synapse.

Feature	electrical	chemical
Transmission time (ms)	~ 0	1
direction	2-way	1-way
neurotransmitter	no	yes

4.2 c.

4.3 see text

4.4 -

4.5 Excitatory synapse: increase PSP. Ion flux at the post-synaptic membrane is transport of additional positively charged ions to the inside and/or more negatively charged ions to the outside. All these events will increase the membrane rest-potential (make it less negative). Probability of firing of post-synaptic neuron will *increase.*

4.6 re-uptake, diffusion from the synaptic cleft, enzymatic breakdown.

4.7 Spatial summation refers to the process where the post-synaptic membrane potential is changed (EPSPs and/or IPSPs) by the action of several action potentials from several neurons (at least two) at the same time. In temporal summation, post-synaptic potential is a function of several action potentials from one neuron over time. Typically, we are faced with spatio-temporal summation, and a receiving neuron will fire if the weighted sum of spatially and temporally summated PSPs exceeds threshold.

4.8 EPSP increases likelihood of firing. IPSP reduces likelihood of firing. Graded potential refers to the local change in membrane potential, that is a typically function of activation of ligand gated channels, such as the acetylcholine receptor.

4.9 a.

4.10 Inhibitory. Membrane potential will typically decrease (become more negative). Note, however, that silent inhibition is a possibility, as well.

4.11 a.

4.12 Place a resistor between two membrane parts. Transmission time is about 0 s.

4.13 See anatomy course.

4.14 Reflexes are specific, predictable, involuntary responses to a particular type of stimulation. Common cutaneous reflexes to test are the abdominal, cremasteric, plantar, anal, and bulbocavernosus. The plantar reflex is performed by striking the sole of the foot with a blunt instrument, such as a tongue depressor or the handle of a reflex hammer. Babinski's reflex is an abnormal[1] plantar reflex, that is present if the great toe flexes toward the top of the foot and the other toes fan out. In people more than 2 years old, the presence of a Babinski's reflex indicates damage to the nerve paths connecting the spinal cord and the brain (the corticospinal tract). Because this tract is right-sided and left-sided, a Babinski's reflex can occur on one side or on both sides.

4.15 Upper motor neuron: ALS (Lou Gehrig's disease) (lower motor neuron involved, as well). Hereditary spastic paresis. Primary lateral sclerosis. Lower motor neuron: ALS (with upper motor neuron), spinal muscular atrophy, poliomyelitis.

4.16 See table B.3.

Table B.3 Differences in clinical examination between an upper (UMN) and lower motor neuron (LMN) disease.

Feature	UMN	LMN
reflex	increased	decreased to absent
atrophy	mild to moderate	severe
muscle tone	increases	decreased
strength	reduced (+ less control)	reduced
pathological reflexes	yes	no

4.17 See text.

[1] Normal in younger children, but abnormal after the age of 2.

Problems of Chapter 5

5.1 The complex of the hair cells, that are slightly deformed by the movement of the endolymph, can be viewed as shape convertors: fluid movement induces mechanical deformation of cells; auxiliary energy is not needed. For the generation of action potentials, however, energy is needed, in particular for the opening and closing of ion channels. Therefore, for the transition of movement of hair cells → action potentials, we do need energy. Therefore, we are dealing now with a modifying transducer (and chemical energy is the auxiliary energy source).

As another example, at the retina, there is a conversion from photons (e.m. radiation) to action potentials. Here, as well, chemical energy is needed for the conversion. However, if we look at this process in more detail, the process starts with light interacting with a particular protein (rhodopsin), present in the rods and cones. Energy from impinging light excites the electrons in the 11-cis-retinal subunit and converts it to a different configuration, 11-trans-retinal. This is a molecular conversion to a different 3-dimensional structure, an no auxiliary energy source is needed. We could view this a as a self-generating process, where light induces a molecular change in configuration. However, considering the whole rod, including the rhodopsin, we have a modulating sensor, where chemical energy is needed as an auxiliary energy source for proper sensor operation.

5.2 Additive drift will induce a systematic error if the sign of the additive drift does not change, and variations are relatively slow. However, in some circumstances, additive drift can also fluctuate, yielding non-systematic

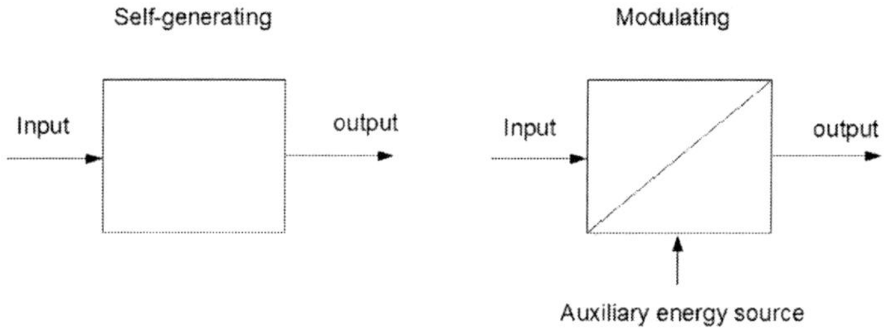

	Input	output	auxiliary
Solar cell	e.m.-radiation	current	-
thermocouple	heat	voltage	-
Heating element	current	heat	-
piezo-electric crystal	force	voltage	-
strain gauge	force	voltage	current
LCD	voltage	radiation	radiation (light)
LED	voltage	radiation	-

Fig. B.3 Left: self generating and right: modulating transducers.

errors. The same applies to sensitivity drift. Therefore, there is not a trivial one-to-one relation between sensitivity drift, additive drift and systematic and non-systematic errors.

5.3 Zero drift: almost all measurement instrumentation, including scales, temperature sensors, humidity sensors, pressure sensors suffer (to some extent) from zero drift. Sensitivity drift is present in almost any sensor, as well, including temperature sensors and pressure sensors. What is relevant in practice, is the magnitude of these effects in relation to the expected values of the measurand and the needed accuracies.

5.4 x_g=86.4 ±4.7
μ_a=187±19.9
μ_b=76.4 ±4.7.

b.) Measurement A has a total error of about 100, measurement B of -10. The inaccuracy of measurement A is approximately 116%, of measurement B about -12%. (cf Eqn. 5.2).

Measurement B is more precise, since it fluctuates less around its value. The sensitivity of measurement A is about 2, of measurement B about 0.88 (you can make a graph of the standard versus the measurement A and B). For measurement B, the sensitivity seems independent of the input signal, while for meaurement B, there are some fluctuations in K. This should be visible in your graph, as well.

5.5 $r \sim 4.4$ mm.

5.6 For the voltage, using a current of 5 mA, use (there are two hands, left and right)

$$V = \left[\frac{1}{10}(R_{\text{left}} + R_{\text{right}}) + 200\right] 5 \cdot 10^{-3} \tag{2.1}$$

which yields

Table B.4 Voltage differences for the several conditions.

condition	voltage difference
dry skin	101 V
wet skin	51 V
with gel	11 V
pierced	1 V

5.7 The half cell potential is a voltage difference that will always arise at the interface of two different conducting media. The *standard* half-cell potential is defined as the voltage difference with reference to the hydrogen electrode. At the electrode-skin interface, therefore, there will always be a voltage difference. If this voltage would remain constant during all conditions, this would pose no problems. The overpotential describes the voltage changes due to current through the interface. In general, the overpotential is a function of the current strength, direction and frequency. The smaller these influence the function of the electrode, the more reliable measurements are possible. As mentioned in the text, the overpotential can be divided into ohmic effects, the concentration overpotential and the activation overpotential.

5.8 The half cell potential for the Ag-AgCl electrode is given by Eqn. 5.15. Substitution of values for the unknowns yields $E_{\text{hc}} = 0.2225 - 0.0256 \ln[a_{Cl^-}]$ Volt. For a typical Cl^- activity at the skin surface, the voltage is approximately 0.349 Volt.

5.9 The overpotential is the voltage change that arises due to polarization effects. The size of the potential change is dictated by the frequency of the current, the current strength, the electrode material and the ion concentration in the electrolyte.

5.10 The use of non-polarizable electrodes, abrasion of the skin with sand paper (which removes part of the stratum corneum), fixation of the electrode and the use of amplifiers with a high input impedance.

5.11 The impedance will be larger, polarization effects will increase, movement artifacts will increase and the measurement will be more noisy.

5.12 For the measured voltage in situation 1, we find that

$$V_1 = \frac{R_1}{R_1 + R_s} \cdot V_{\text{source}} \tag{2.1}$$

with $R_1 = 5\ \text{k}\Omega$, R_s the source impedance and V_{source} the source voltage. A similar expression exists in situation 2. Combining these two equations results in

$$R_s = \frac{R_1 R_2 \cdot (V_2 - V_1)}{V_1 R_2 - V_2 R_1}. \tag{2.2}$$

Inserting the correct values, we find for the source impedance $Rs = 40\ \Omega$.

Problems of Chapter 6

6.1 For the amount of energy delivered, it holds that

$$P = (I^2 \cdot R_{\text{skin}}) \cdot \Delta t \tag{2.3}$$

Using $R_{\text{skin}} = 50\ \text{k}\Omega$, we find 80 mJ.

6.2 see text

6.3 3-6 see text

6.7 The conductivity, σ is defined as $\sigma = l/RA$ with R the total resistance, l, the length and A the cross-sectional area of the specimen. The current source is placed in a homogeneous volume conductor. Estimate the voltage across a spherical layer, as indicated in Fig. B.4.

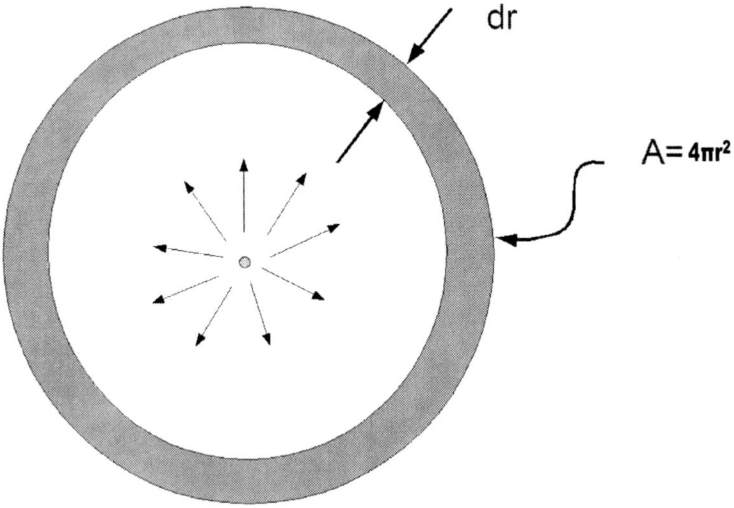

Fig. B.4 Slice of the sphere, with thickness dr.

For the change in voltage it holds that

$$dV = IdR = I\frac{dr}{\sigma A} = I\frac{dr}{\sigma 4\pi r^2} \quad (2.1)$$

with dr the change in radius and R the resistance. Integration yields Eqn. 6.1.

Problems of Chapter 7

7.1 c

7.2 b

7.3 a) see text.
b) myasthenia gravis, Lambert-Eaton myasthenic syndrome, botulism
c) sodium
d) at the pre-synaptic nerve ending (interferes with the excretion of ACh).

7.4 b

7.5 c

7.6 a

7.7 Curare is one of the non-depolarizing muscle relaxants. It competes with acetylcholine at the NMJ to bind with the ACh-receptor; however, it does not activate the receptor. It therefore results in muscle paralysis. Non-depolarizing muscle relaxants as curare also act on presynaptic receptors, as well, causing an inhibition of the release of ACh.

Muscle relaxants are often used during surgery to provide muscle relaxation and allow mechanical ventilation. There is a differential sensitivity for the various muscles to curare, the diaphragm being the least sensitive. This is why patients undergoing surgery may sometimes hiccup, as an early sign that the muscle relaxants are wearing off.

Problems of Chapter 8

8.1 c.

8.2 A montage is a particular electrode configuration that is used to measure a voltage difference across the scalp.

8.3 10-100 µV.

8.4 -

8.5 The phenomenon that the alpha rhythm is suppressed by opening of the eyes. This is also known as "reactivity". If this is reduced or even absent, it might indicate a global brain dysfunction, as e.g. encountered in some forms of Alzheimer's disease.

8.6 Delta waves are between 0.5 and 4 Hz. They are a normal phenomenon during sleep.

8.7 cerebral infarction or hemorrhage, trauma, unilateral seizures, tumor.

8.8 –

8.9 The EEG observed in patients suffering from epilepsy during moments there are no clinical signs of seizure activity. Oftentimes, this interictal EEG will show some abnormalities, e.g. an isolated peak or sharp waves.

8.10 peaks, sharp waves, focal slowing. They indicate short-lasting moments of abnormal synchrony of relatively small parts of neuronal tissue. Both the short duration and the limited amount of neuronal tissue involved make that clinical signs will not be perceived by the patient or noted by an observer at these moments.

Problems of Chapter 9

9.1 An analog filter assures that the Nyquist criterion is met. If this filter is not applied, and frequencies in the analog signal are present that are sampled with a sampling frequency less than twice their frequency, these frequencies will be "aliased" to artefactual lower frequencies in the digitized signal. You can easily verify this statement by a simulation in Matlab.

9.2 Four epochs of 2 s each are needed, at least, as shown in Table 9.1.

9.3 You need 200/0.1=2000 discretizations. For this, you will need at least an 11 bit AD-convertor, since $2^{11} = 1024$.

Problems of Chapter 10

10.1 -

10.2 Slowing of the heart rate.

10.3 The conduction velocities are 30-90, 10-20 and 0.3-6 m/s for A-, B-, and C-fibers, respectively.

10.4 -

10.5

10.6 Current thresholds are 0.02-0.2, 0.04-0.6 and > 2 mA for A, B and C-fibers.

Problems of Chapter 11

11.1 -

11.2 a) This remains 30%. the likelihood does not change using a test with a positive likelihood ratio of 1.
b) The positive likelihood ratio is now 9. Use Bayes to find P(D) $\sim$ 0.79.

11.3 In the Monte Hall three door problem the likelihood that the price is now behind door C has increased to 2/3. So switching to door C doubles the likelihood of winning the price (the chance of winning the price remaining at door B remains 1/3).

11.4 For the likelihood ratio, we find LR=115; specificity approximately 99.3%

11.5 The sensitivity of a routine EEG in the diagnosis of epilepsy is approximately 30-50%. This implies that a normal EEG does not strongly reduce the prior probabilities. Therefore, the correct answer is b.

11.6 -

Index

Printed in the United States
154245LV00001B/41/P

9 783540 698890